T0135504

# Tomographic determination of seismic velocity models with kinematic wavefield attributes

---

# Tomographische Bestimmung seismischer Geschwindigkeitsmodelle mit kinematischen Wellenfeldattributen

Zur Erlangung des akademischen Grades eines

DOKTORS DER NATURWISSENSCHAFTEN

von der Fakultät für Physik der

Universität Karlsruhe (TH)

genehmigte

DISSERTATION

von

Dipl.-Geophys. Eric Duveneck

aus

Bremen

Tag der mündlichen Prüfung:             25. Juni 2004

Referent:                                   Prof. Dr. Peter Hubral

Korreferent:                         Prof. Dr. Friedemann Wenzel

Bibliografische Information Der Deutschen Bibliothek

Die Deutsche Bibliothek verzeichnet diese Publikation in der Deutschen
Nationalbibliografie; detaillierte bibliografische Daten sind im Internet über
http://dnb.ddb.de abrufbar.

ISBN 3-8325-0647-0

Logos Verlag Berlin
Comeniushof, Gubener Str. 47,
10243 Berlin
Tel.: +49 030 42 85 10 90
Fax: +49 030 42 85 10 92
INTERNET: http://www.logos-verlag.de

# Abstract

For the transformation of recorded seismic reflection data into a structural image of the subsurface by depth migration, a seismic velocity model is required. A commonly used tool for the construction of such velocity models in laterally inhomogeneous media is reflection tomography. One of the drawbacks of that method is, however, that it requires picking of reflection events in the seismic prestack data to provide the traveltime information for the tomographic inversion. This picking is extremely time-consuming, especially in 3D seismic data, and can become difficult or even impossible if the signal-to-noise ratio in the data is low.

In this thesis, a new tomographic inversion method for the determination of smooth, isotropic velocity models is presented that makes use of traveltime information in the form of kinematic wavefield attributes. These attributes are the coefficients of second-order traveltime approximations in the midpoint and offset coordinates and can be extracted from the seismic prestack data by means of a coherence analysis, e. g., with the common-reflection-surface (CRS) stack. The required input data for the tomographic inversion are taken from the CRS stack results at a number of pick locations in the CRS-stacked simulated zero-offset section. Picking is further simplified by the fact that, due to the used model parametrization in terms of a smooth velocity distribution and a number of isolated reflection points, the pick locations do not need to follow continuous, interpreted horizons in the stacked section, but are independent of each other and may be placed on locally coherent events.

The attributes used in the tomographic inversion can be interpreted in terms of the second-order traveltimes of hypothetical emerging wavefronts due to a point source at the normal-incidence point of the respective zero-offset ray, so-called NIP wavefronts. During the inversion process a model is found that minimizes the misfit between these data and the corresponding quantities modeled by dynamic ray tracing along the associated central (normal) rays.

In the thesis, the complete theory of the method, as well as practical applications to synthetic and real seismic data are presented. Starting with an overview of the required ray theory results and the CRS stack method, the general concept of the new tomographic inversion approach based on kinematic wavefield attributes is developed. Practical and implementation-related aspects of the method are then discussed for the case of 1D, 2D, and 3D tomographic inversion. Finally, the entire process of deriving a velocity model from seismic data is demonstrated on a synthetic and on a real 2D seismic dataset. Beginning with the prestack data, the CRS stack is performed, the required input data for the inversion are picked and extracted from the CRS stack results, and the tomographic inversion itself is applied. The consistency of the resulting velocity models with the respective seismic prestack data is verified by the application of prestack depth migration.

# Zusammenfassung

Die vorliegende Arbeit ist mit Ausnahme dieser Zusammenfassung in englischer Sprache geschrieben. Im Folgenden sind die wesentlichen Aspekte der Arbeit kapitelweise kurz auf Deutsch zusammengefasst. Englischsprachige Fachbegriffe sind dabei *kursiv* gesetzt.

## Einleitung

Die Reflexionsseismik spielt eine wichtige Rolle bei der globalen Exploration nach Kohlenwasserstoffreserven im Erduntergrund. Das Grundprinzip der reflexionsseismischen Methode besteht darin, mithilfe kontrollierter seismischer Quellen elastische Energie in den Untergrund einzubringen und das seismische Wellenfeld zu messen und auszuwerten, das wieder die Erdoberfläche erreicht nachdem es an Diskontinuitäten der elastischen Eigenschaften des Erduntergrundes reflektiert wurde.

Sowohl die Laufzeiten als auch die Amplituden der gemessenen Wellen enthalten Informationen, die es erlauben, ein detailliertes Abbild dieser Diskontinuitäten und damit der geologischen Strukturen im Untergrund zu erhalten, sowie möglicherweise quantitative Aussagen über die zugehörigen elastischen Eigenschaften zu machen. Strukturen, die für die Kohlenwasserstoffexploration von Interesse sind, befinden sich normalerweise in Tiefen von bis zu 5 km.

Seismische Messungen werden sowohl an Land als auch auf dem Meer durchgeführt. Sie bestehen im Allgemeinen aus einer Vielzahl von Einzelexperimenten, bei denen jeweils eine seismische Quelle und eine große Anzahl seismischer Empfänger mit verschiedenen Entfernungen von der Quelle (*offsets*) verwendet werden (Abbildung 1.1). Je nachdem, ob ein zwei- oder dreidimensionales Abbild des Erduntergrundes angestrebt wird, sind die Empfänger auf einer Linie mit der Quelle oder flächenhaft angeordnet. Man spricht dann von 2D bzw. 3D Akquisition. In der Landseismik werden als seismische Quellen Explosionen oder seismische Vibrationsquellen verwendet, während in der marinen Seismik sogenannte *air guns*, die einen Druckluftpuls ins Wasser bringen, benutzt werden. Das an den jeweiligen Empfängern in Form der Partikelverschiebung (Land) oder von Druckvariationen (Wasser) gemessene Wellenfeld wird in Form von digitalen Zeitreihen aufgezeichnet.

Durch die Überlappung der Messanordnungen der verschiedenen Einzelexperimente ergibt sich ein sogenannter mehrfach überdeckter seismischer Datensatz, der redundante Informationen über die zu untersuchenden Untergrundstrukturen enthält. Diese Redundanz spielt im weiteren Verlauf der seismischen Datenbearbeitung und Auswertung eine wichtige Rolle. Die Datenbearbeitung zielt darauf ab, nicht nutzbare Wellentypen und Rauschen in den gemessenen Daten zu unterdrücken und die für die Reflexionsseismik relevanten Signale (Primärreflexionen) hervorzuheben. Dabei wird unter anderem häufig eine Summation (Stapelung) von seismischen Signalen mit verschiedenen *offsets* aber gleichem Mittelpunkt (*midpoint*) zwischen Quelle und Empfänger vorgenommen, um das Datenvolumen zu reduzieren und das Signal/Rauschen (S/R) Verhältnis zu verbessern. Das Resultat kann dann näherungsweise als Simulation einer seismischen Sektion angesehen werden, wie man sie bei Verwendung von koinzidenten Quell- und Empfängerlokationen (*zero-offset*) erhalten würde.

Eines der Hauptziele der Reflexionsseismik ist es, aus den gemessenen seismischen Daten ein strukturelles Tiefenabbild des Untergrundes zu erhalten. Dazu wird eine sogenannte Tiefenmigration durchgeführt. Während des Migrationsprozesses werden die Effekte der Wellenausbreitung im Untergrund rechnerisch rückgängig gemacht und so die gemessenen Reflexionsereignisse in Reflektorabbilder im Untergrund transformiert. Entsprechend basieren alle gängigen Migrationsmethoden auf der Wellengleichung. Sie beinhalten normalerweise (konzeptionell) zwei Schritte: die Wellenfeldextrapolation in den Untergrund und die Anwendung einer Abbildungsbedingung, um aus dem extrapolierten Wellenfeld das Tiefenabbild zu erhalten.

Aufgrund der Tatsache, dass der Migrationsprozess auf einer Wellenfeldextrapolation beruht, wird zur Durchführung einer Tiefenmigration ein Modell der Ausbreitungsgeschwindigkeiten der seismischen Wellen im Untergrund benötigt. Ein solches Modell ist im Allgemeinen zunächst nicht bekannt und muss, unter Zuhilfenahme von möglicherweise vorhandenen geologischen Vorinformationen und Bohrlochmessungen, aus den seismischen Daten selbst konstruiert werden. Zu diesem Zweck wird Gebrauch von der bereits erwähnten Redundanz in mehrfach überdeckten seismischen Daten gemacht. Diese Daten enthalten zwar nicht genügend Information, um die wahre Verteilung der seismischen Geschwindigkeiten im Untergrund zu rekonstruieren, es lassen sich aber normalerweise Geschwindigkeitsmodelle bestimmen, die konsistent mit den seismischen Daten sind. Solche für die Tiefenmigration optimalen Modelle entsprechen in ihren groben Strukturen der wahren Geschwindigkeitsverteilung und werden Makrogeschwindigkeitsmodelle genannt.

Zur Bestimmung von Geschwindigkeitsmodellen sind eine Reihe verschiedener Methoden in Gebrauch, die alle auf dem Kriterium der Konsistenz des Modells mit den Daten basieren, sich aber darin unterscheiden, wie diese Konsistenz gemessen wird, wie das Modell parametrisiert ist und wie Inkonsistenzen in Verbesserungen des Modells umgesetzt werden. Aufgrund der Nichtlinearität des Problems arbeiten Methoden zur Bestimmung von Geschwindigkeitsmodellen normalerweise iterativ.

Eine gängige Methode zur Modellbestimmung in lateral inhomogenen Medien ist die Reflexionstomographie. In reflexionstomographischen Methoden (auch Laufzeitinversion genannt) werden in den seismischen Daten Reflexionsereignisse ausgewählt ("gepickt") und die zugehörigen Laufzeiten mit solchen verglichen, die in einem vorgegebenen Modell, bestehend aus Reflektoren und einer Geschwindigkeitsverteilung, vorwärts berechnet wurden. Laufzeitdifferenzen werden dann durch eine tomographische Inversion in eine Verbesserung des Modells umgerechnet. Dieser Prozess wird solange wiederholt, bis der Laufzeitfehler ausreichend reduziert ist. Das Picken der für die Reflexionstomographie benötigten Reflexionsereignisse in den ungestapelten seismischen

Daten ist aufgrund der großen Datenmenge allerdings sehr aufwändig. Normalerweise wird es interaktiv durchgeführt, wobei die gepickten Ereignisse auf kontinuierlichen interpretierten Horizonten in den seismischen Daten liegen müssen. Besonders bei niedrigem S/R Verhältnis sind Reflexionsereignisse in den ungestapelten Daten oft nur schwer zu identifizieren, was das Picken zusätzlich erschwert.

In der vorliegenden Arbeit wird eine neue, alternative reflexionstomographische Methode vorgestellt, die auf einer sehr viel effizienteren Gewinnung der benötigten Laufzeitinformationen beruht. Dazu werden Laufzeitapproximationen zweiter Ordnung in den *midpoint-* und *offset-*Koordinaten verwendet. Die Laufzeitinformation ist dann in den zugehörigen Koeffizienten der Approximation enthalten. Diese Koeffizienten, auch als kinematische Wellenfeldattribute bezeichnet, lassen sich zum Beispiel mittels der *common-reflection-surface (CRS) stack* Methode durch Kohärenzanalysen in den ungestapelten seismischen Daten gewinnen. Obwohl der Gebrauch von Laufzeitapproximationen die Anwendbarkeit der Methode in Fällen starker lateraler Geschwindigkeitsvariationen einschränkt, führt er zu einer Reihe von klaren Vorteilen in der praktischen Anwendung.

Die für die tomographische Inversion benötigten kinematischen Wellenfeldattribute werden an ausgewählten Picklokationen aus den *CRS-stack* Ergebnissen extrahiert. Das Picken kann in der aus dem *CRS stack* resultierenden gestapelten, simulierten *zero-offset* Sektion durchgeführt werden, die ein erheblich verbessertes S/R Verhältnis im Vergleich zu den ungestapelten seismischen Daten bei gleichzeitig stark reduzierter Datenmenge aufweist. Aufgrund der speziellen bei der tomographischen Inversion verwendeten Modellparametrisierung müssen die Picklokationen zudem nicht auf kontinuierlichen Horizonten liegen, sondern können sich unabhängig voneinander auf lokal kohärenten Ereignissen befinden. Insgesamt wird das Picken der für die tomographische Inversion benötigten Laufzeitinformationen, besonders im Fall von Daten mit niedrigem S/R Verhältnis, erheblich vereinfacht.

## Seismische Strahlentheorie

Die in dieser Arbeit vorgestellte tomographische Methode zur Bestimmung seismischer Geschwindigkeitsmodelle macht umfassenden Gebrauch von verschiedenen Elementen der seismischen Strahlentheorie. Die seismische Strahlentheorie ermöglicht eine effiziente Beschreibung der Ausbreitung hochfrequenter seismischer Wellenfelder in inhomogenen Medien. Sie ist anwendbar solange die dominierenden Wellenlängen des zu beschreibenden Wellenfeldes deutlich kleiner sind als die charakteristischen Längenskalen der Inhomogenitäten des durchlaufenen Mediums.

Für die Formulierung der tomographischen Inversion werden nur die kinematischen Aspekte der seismischen Strahlentheorie benötigt, daher werden nur diese im Folgenden genauer beschrieben. Eine entscheidende Rolle spielt dabei die Eikonalgleichung, eine nichtlineare, partielle Differentialgleichung erster Ordnung, die sich aus der Wellengleichung mittels eines entsprechenden Lösungsansatzes unter Annahme hoher Frequenzen gewinnen lässt. Sie beschreibt die Ausbreitung von möglichen Unstetigkeitsflächen des Wellenfeldes, bzw. von Flächen gleicher Phase, die als Wellenfronten angesehen werden können. Die Eikonalgleichung lässt sich mithilfe der Methode der Charakteristiken lösen, die auf ein System gewöhnlicher Differentialgleichungen in den Ortskoordinaten und den Komponenten des zugehörigen *slowness-*Vektors führt. Die Lösungen dieses

Systems lassen sich physikalisch als Strahlen interpretieren, weshalb man auch vom *ray-tracing* System spricht. Im hier betrachteten isotropen Fall stehen die Strahlen senkrecht zu den zugehörigen Wellenfronten.

Entwickelt man das *ray-tracing* System in eine Taylor-Reihe bis zur ersten Ordnung in den Ortskoordinaten und *slowness*-Komponenten um einen bereits bestimmten Strahl (den Referenzstrahl), so erhält man ein lineares System gewöhnlicher Differentialgleichungen in den Variationen der Orts- und *slowness*-Komponenten, das so genannte paraxiale *ray-tracing* System. Dieses beschreibt Strahlen in der paraxialen Umgebung um den Referenzstrahl. Aufgrund seiner Linearität lassen sich allgemeine Lösungen des paraxialen *ray-tracing* Systems durch eine Fundamental- oder Propagatormatrix darstellen. Das dem paraxialen *ray-tracing* System formal entsprechende dynamische *ray-tracing* System erlaubt unter anderem, Laufzeitapproximationen bis zur zweiten Ordnung von vorgegebenen Wellen (spezifiziert durch ihre Anfangsbedingungen) in der Umgebung des betrachteten Referenzstrahls zu berechnen. Insbesondere können die Elemente der Propagatormatrix als Lösungen des dynamischen *ray-tracing* Systems für zwei spezifische Wellen aufgefasst werden. Von diesen ist für den weiteren Verlauf der Arbeit die aus einer Punktquelle resultierende Welle von besonderer Bedeutung. Aus den zugehörigen Elementen der Propagatormatrix lassen sich die zweiten Laufzeitableitungen dieser Welle um den Referenzstrahl berechnen.

Desweiteren werden die Ergebnisse der Strahlen-Perturbationstheorie behandelt. Sie erlauben es, die Auswirkungen von Perturbationen der Anfangsbedingungen von Strahlen, aber auch der Geschwindigkeitsverteilung im Hintergrundmedium, auf den Strahlverlauf zu bestimmen. Außerdem lassen sich die Effekte solcher Perturbationen auf die zugehörige Propagatormatrix, das heißt auf den Verlauf von paraxialen Strahlen in der Umgebung des Referenzstrahls, bis zur ersten Ordnung berechnen. Die Ergebnisse der Strahlen-Perturbationstheorie spielen für die Formulierung der auf Laufzeitapproximationen basierenden tomographischen Inversion eine wichtige Rolle.

Alle bisher genannten Aspekte der Strahlentheorie werden in allgemeinen krummlinigen orthogonalen Koordinatensystemen formuliert. In der tomographischen Inversion werden zwei Koordinatensysteme verwendet: strahlzentrierte Koordinaten und globale kartesische Koordinaten. In strahlzentrierten Koordinaten (Abbildung 2.1) vereinfachen sich viele der zuvor hergeleiteten Ergebnisse. Bei Kenntnis der lokalen Geschwindigkeit lassen sich die zweiten räumlichen Laufzeitableitungen in strahlzentrierten Koordinaten auch in Wellenfrontkrümmungen umrechnen.

## Die Common-Reflection-Surface Stack Methode

Wie bereits erwähnt wird während der seismischen Datenbearbeitung häufig eine Stapelung von Spuren mit koinzidentem *midpoint* aber verschiedenen *offsets* (in einem sogenannten *common-midpoint* oder *CMP gather*) durchgeführt, um das S/R Verhältnis zu verbessern und die Datenmenge zu reduzieren. Zu diesem Zweck werden die mehrfach überdeckten seismischen Daten zunächst bezüglich ihrer *midpoint*- und *offset*-Koordinaten sortiert. Als Summationstrajektorien für den sogenannten *CMP stack* wird eine Laufzeitapproximation zweiter Ordnung in der *offset* Koordinate verwendet. Für gegebene *zero-offset* Laufzeit enthält diese Laufzeitapproximation (im 2D Fall) einen zu bestimmenden Parameter, die sogenannte Stapelgeschwindigkeit. Optimale

Werte der Stapelgeschwindigkeit werden mittels einer Kohärenzanalyse im *CMP gather*, einer so genannten Geschwindigkeitsanalyse, bestimmt (Abbildung 3.3).

Dieser Prozess lässt sich verallgemeinern, indem (im 2D Fall) anstatt der Summationstrajektorie in *offset* Richtung eine Summationsfläche verwendet wird, die sich sowohl in die *offset* als auch in die *midpoint* Richtung erstreckt (Abbildung 3.4). Das S/R Verhältnis in der resultierenden Stapelsektion lässt sich damit weiter verbessern. Dies ist das Konzept der *common-reflection-surface (CRS) stack* Methode. Die dazu verwendete Laufzeitapproximation zweiter Ordnung in der *offset* und der *midpoint* Koordinate hat im 2D Fall drei zu bestimmende Koeffizienten, die sogenannten kinematischen Wellenfeldattribute. Sie werden für jede Lokation und *zero-offset* Laufzeit der zu simulierenden *zero-offset* Sektion (oder *CRS-stack* Sektion) automatisch mittels einer Kohärenzanalyse bestimmt. Dazu werden jeweils die kinematischen Wellenfeldattribute so lange variiert, bis die Kohärenz entlang der Summationsfläche maximal wird. Für jede *zero-offset* Lokation und Laufzeit erhält man so einen Satz von drei optimalen kinematischen Wellenfeldattributen.

Die mit einem Reflexionsereignis in der *CRS-stack* Sektion assoziierten kinematischen Wellenfeldattribute repräsentieren zunächst nichts anderes als die ersten und zweiten Ableitungen der Reflexionslaufzeit in *offset* und *midpoint* Richtung (wobei die erste Ableitung in *offset* Richtung null ist). Es lässt sich aber zeigen, dass sich diese Attribute mit zwei hypothetischen, an der betrachteten *midpoint* Lokation auftauchenden Wellenfronten (Abbildung 3.5) in Verbindung bringen lassen. Beide hypothetischen Wellen haben ihren Ursprung am Reflexionspunkt des zugehörigen *zero-offset* Strahls auf dem Reflektor, dem sogenannten *normal-incidence point*, oder NIP. Eine dieser Wellen, die Normal-Welle ergibt sich aufgrund eines (hypothetischen) explodierenden Reflektorelements am NIP, mit einer dem wahren Reflektor entsprechenden lokalen Krümmung. Die andere Welle ergibt sich aufgrund einer (hypothetischen) Punktquelle am NIP und wird NIP-Welle genannt. Die kinematischen Wellenfeldattribute beschreiben dann die erste und zweite Laufzeitableitung der auftauchenden hypothetischen Normal- und NIP-Wellen bezüglich der räumlichen Koordinate entlang des seismischen Profils (die erste Ableitung ist für beide Wellen identisch). Ist die lokale oberflächennahe Geschwindigkeit bekannt, lassen sich diese ersten und zweiten Laufzeitableitungen umrechnen in Wellenfrontkrümmungen der beiden hypothetischen Wellen sowie den zugehörigen Auftauchwinkel des Normalstrahls (des am NIP senkrecht auf dem Reflektor stehenden Strahls).

Im 3D Fall werden die *offset*- und *midpoint*-Koordinaten jeweils durch zweikomponentige Vektoren dargestellt. Die beim *CRS stack* verwendete Laufzeitapproximation hängt dann von bis zu acht kinematischen Wellenfeldattributen ab, wobei jeweils drei Attribute die symmetrischen Matrizen der zweiten Ableitungen der Normal- und NIP-Wellen-Laufzeiten definieren und zwei Attribute die ersten Laufzeitableitungen repräsentieren.

Der *CRS stack* kann als Methode angesehen werden, Laufzeitinformationen in Form der kinematischen Wellenfeldattribute aus den seismischen Daten zu extrahieren. Für die im Folgenden beschriebene tomographische Inversion werden dazu nur diejenigen Attribute verwendet, die auftauchende hypothetische NIP-Wellen beschreiben.

# Tomographische Inversion mit kinematischen Wellenfeldattributen

In einem mit den Daten konsistenten Geschwindigkeitsmodell werden Reflexionsereignisse in den seismischen Daten, die zu einem gemeinsamen Reflexionspunkt im wahren Untergrund gehören, durch die Anwendung einer Tiefenmigration an einen gemeinsamen Punkt migriert. Die zugehörigen Strahlsegmente (Abbildung 4.2a) sind geometrisch identisch mit den Strahlen einer hypothetischen NIP-Welle (Abbildung 4.2b). Daraus ergibt sich, dass in einem konsistenten Geschwindigkeitsmodell die NIP-Welle, wenn sie entlang des Normalstrahls in den Untergrund propagiert wird, bei der Laufzeit null am NIP fokussieren muss (Abbildung 4.2c). Dieses Kriterium erlaubt es, mithilfe der kinematischen Wellenfeldattribute, die die NIP-Wellen-Laufzeit beschreiben, ein konsistentes Geschwindigkeitsmodell zu bestimmen.

Methoden, die dieses Prinzip verwenden, um ein geschichtetes Modell zu erstellen, sind seit längerem bekannt. Sie benötigen die zugehörigen kinematischen Wellenfeldattribute entlang kontinuierlicher interpretierter Reflektorhorizonte in den seismischen Daten, die den Diskontinuitäten oder Schichtgrenzen im Modell zugeordnet sind. Solche über weite Bereiche kontinuierliche Reflexionsereignisse sind allerdings nicht immer vorhanden oder können nicht überall verläßlich identifiziert werden, so dass schichtbasierte Methoden oft nur beschränkt anwendbar sind.

Im Folgenden wird eine neue, flexiblere, auf einem tomographischen Ansatz beruhende Methode zur Bestimmung isotroper seismischer Geschwindigkeitsmodelle aus kinematischen Wellenfeldattributen vorgestellt. Sie macht Gebrauch von einer glatten Modellbeschreibung ohne Diskontinuitäten und damit ohne Schichtgrenzen. Das Grundprinzip der Methode besteht darin, durch tomographische Inversion ein Modell zu finden, in dem die durch kinematische Wellenfeldattribute beschriebenen auftauchenden NIP-Wellen korrekt modelliert werden, das heißt, die Abweichung zwischen den aus den seismischen Daten extrahierten und den entsprechenden vorwärts modellierten Attributen minimiert wird.

Jeder eine NIP-Welle repräsentierende Datenpunkt besteht dabei aus der Einweglaufzeit (der halben *zero-offset* Laufzeit) und dem Auftauchpunkt des zugehörigen Normalstrahls (der betrachteten Lokation in der simulierten *zero-offset* Sektion), sowie den ersten und zweiten Laufzeitableitungen der jeweiligen NIP-Welle (den entsprechenden kinematischen Wellenfeldattributen). Die benötigten Datenkomponenten werden an ausgewählten Picklokationen aus den Ergebnissen des *CRS stack* extrahiert. Das zu bestimmende Modell besteht aus zwei Teilen: dem Geschwindigkeitsmodell selbst und den zu den Daten gehörigen Reflexionspunkten (NIPs). Das Geschwindigkeitsmodell wird durch B-Splines, das heißt durch glatte, lokalisierte Spline-Basisfunktionen, beschrieben. Die zugehörigen B-Spline Koeffizienten stellen dabei die zu bestimmenden Modellparameter dar. Jeder NIP wird repräsentiert durch seine Ortskoordinaten, sowie durch die lokale Reflektororientierung, darstellbar in 2D durch einen, in 3D durch zwei Parameter. Die zu einer auftauchenden hypothetischen NIP-Welle gehörenden Modell- und Datenkomponenten sind in Abbildung 4.3 dargestellt.

Bei gegebenen Modellparametern lassen sich zu jedem Datenpunkt die entsprechenden vorwärts modellierten Größen durch dynamisches *ray tracing* entlang des jeweiligen, durch die NIP-Modellparameter definierten, Normalstrahls berechnen. Die zweiten Laufzeitableitungen einer NIP-Welle können dabei einfach mithilfe der zugehörigen Elemente der Propagatormatrix bestimmt werden, während der Auftauchpunkt und die Laufzeit entlang des Normalstrahls durch gewöhnliches *ray tracing* zu berechnen sind. Die erste horizontale Ableitung der NIP-Wellenlaufzeit ist identisch mit der aus dem *ray tracing* resultierenden horizontalen *slowness*-Komponente.

Das zu lösende Inversionsproblem besteht wie bereits erwähnt darin, ein Modell zu finden, in dem die Abweichung zwischen den gegebenen Datenpunkten und den entsprechenden vorwärts modellierten Größen minimiert wird. Dies stellt ein nichtlineares Optimierungsproblem dar. Es lässt sich iterativ im Sinne der kleinsten Fehlerquadrate lösen, indem es während jeder Iteration lokal linearisiert wird. Das jeweils resultierende zu lösende lineare Gleichungssystem enthält die Fréchet-Ableitungen, das heißt, die partiellen Ableitungen des Modellierungsoperators (des dynamischen *ray tracing*) bezüglich der Modellparameter. Diese Ableitungen können während des *ray tracing* mit den Methoden der Strahlen-Perturbationstheorie berechnet werden.

Wie bei fast allen tomographischen Problemen, enthalten auch im vorliegenden Fall die Daten nicht genügend Informationen, um die Modellparameter eindeutig zu bestimmen. Um das Inversionsproblem zu regularisieren, müssen zusätzliche Bedingungen an die Modellparameter gestellt werden. Eine solche physikalisch plausible Bedingung besteht darin, ein Geschwindigkeitsmodell mit möglichst wenig unnötiger Struktur zu fordern. Das heißt, es wird das einfachste, glatteste Modell gesucht, das die Daten erklärt. Numerisch lässt sich diese Bedingung durch die Minimierung der zweiten räumlichen Ableitungen der Geschwindigkeitsverteilung während der tomographischen Inversion realisieren. Dies führt auf ein erweitertes Matrix-Gleichungssystem. Zusätzlich können eine Reihe anderer Bedingungen an das Modell gestellt werden, zum Beispiel um möglicherweise vorhandene Vorinformationen über die Geschwindigkeitsverteilung zu berücksichtigen.

# Implementierung

Während die tomographische Inversion mit kinematischen Wellenfeldattributen zunächst für den allgemeinen Fall formuliert wurde, wird sie nun spezialisiert auf den 1D Fall (horizontale Reflektoren und eine nur in vertikaler Richtung variierende Geschwindigkeitsverteilung), den 2D Fall (zweidimensionale Geschwindigkeits- und Reflektorstrukturen), sowie den 3D Fall mit beschränkter Azimut-Information (die zweite Ableitung der NIP-Wellenlaufzeit ist nur in einer Azimut-Richtung vorhanden).

Für jeden der drei behandelten Fälle werden die benötigten Daten- und Modellkomponenten, die Struktur der erweiterten tomographischen Matrix, sowie Aspekte der Vorwärtsmodellierung, der Berechnung der Fréchet-Ableitungen und der numerischen Lösung des resultierenden linearen Gleichungssystems diskutiert. Außerdem werden die verschiedenen während der Inversion benötigten Parameter, Gewichtungs- und Skalierungsfaktoren, sowie Aspekte der Regularisierung behandelt. Desweiteren wird der Inversionsalgorithmus für jeden der drei Fälle an einem synthetischen Beispiel demonstriert.

Im 1D Fall besteht jeder Datenpunkt nur aus der Einweglaufzeit und der zweiten Zeitableitung der NIP-Welle. Jeder NIP im Untergrund ist durch seine Tiefe vollständig definiert. Die benötigten Ausdrücke für die Vorwärtsmodellierung und die Berechnung der Fréchet-Ableitungen nehmen im 1D Fall eine besonders einfache Form an. Zur Demonstration der 1D tomographischen Inversion wird ein in einem geschichteten Modell mittels *ray tracing* modellierter *CMP gather* benutzt (Abbildung 5.1). Mit den aus diesem *CMP gather* extrahierten kinematischen Wellenfeldattributen lässt sich ein glattes Geschwindigkeitsmodell rekonstruieren, das dem tatsächlichen, geschichteten Modell sehr nahe kommt (Abbildung 5.2). Insbesondere stimmen die resultierenden NIP-Tiefen sehr gut mit den tatsächlichen Reflektortiefen überein.

Im 2D Fall enthält jeder Datenpunkt vier Komponenten: die Einweglaufzeit und den Auftauchpunkt des Normalstrahls, sowie die erste und die zweite Laufzeitableitung der NIP-Welle. Jeder NIP im Untergrund ist durch zwei räumliche Koordinaten und einen Winkel (die lokale Reflektorneigung) beschrieben (Abbildung 5.7). Die Vorwärtsmodellierung durch dynamisches *ray tracing*, sowie die Anwendung der Strahlen-Perturbationstheorie zur Berechnung der Fréchet-Ableitungen wird im 2D Fall in strahlzentrierten Koordinaten durchgeführt.

Im 3D Fall mit beschränkter Azimut-Information besteht jeder Datenpunkt aus sechs Komponenten: der Einweglaufzeit und zwei Koordinaten des Auftauchpunkts des Normalstrahls, zwei erste Ableitungen der NIP-Wellenlaufzeit, sowie eine zweite Ableitung der NIP-Wellenlaufzeit in einer vorgegebenen Azimut-Richtung (Abbildung 5.13). Jeder NIP im Untergrund ist durch seine drei räumlichen Koordinaten sowie durch zwei horizontale Komponenten des zum Reflektor lokal senkrechten Einheitsvektors bestimmt.

Zur Demonstration des Inversionsalgorithmus, sowohl in 2D als auch in 3D, wird hier zunächst jeweils eine Anwendung auf perfekte, mit dynamischem *ray tracing* in einer glatten 2D bzw. 3D Geschwindigkeitsverteilung (Abbildungen 5.8 und 5.14) modellierte, Datenpunkte gezeigt. Es sind also keine seismischen Daten involviert. In beiden Fällen lässt sich das Geschwindigkeitsmodell sehr gut rekonstruieren (Abbildungen 5.11, 5.19 und 5.20). Unterschiede finden sich lediglich im untersten Teil des Modells, durch den keine (oder nur wenige) Strahlen verlaufen. Der Fehler in den rekonstruierten NIP-Lokationen ist vernachlässigbar klein (Abbildungen 5.12a und b, Abbildung 5.21).

Die Robustheit der Methode lässt sich untersuchen, indem die bei der Inversion verwendeten Daten mit Rauschen versehen werden. Dies wird für die 2D tomographische Inversion mit verschiedenen Realisationen des Rauschens wiederholt durchgeführt. Obwohl, wie zu erwarten, die rekonstruierten NIP-Lokationen eine gewisse Streuung aufweisen, kann die Reflektorstruktur insgesamt in allen Fällen gut rekonstruiert werden (Abbildung 5.12e).

# Anwendungen

Um in der Praxis mittels der beschriebenen tomographischen Inversion ein Geschwindigkeitsmodell aus mehrfach überdeckten seismischen Daten zu erhalten, muss zunächst der *CRS stack* auf die Daten angewendet werden. In der resultierenden simulierten *zero-offset* Sektion werden dann eine Reihe von Punkten auf Reflexionsereignissen gepickt und die zugehörigen für die Inversion benötigten Attribute aus den *CRS-stack* Ergebnissen extrahiert. Bei diesen vorbereitenden Datenbearbeitungsschritten sind eine Reihe praktischer Aspekte zu beachten.

Die Verlässlichkeit der mit dem *CRS stack* aus den seismischen Daten gewonnenen kinematischen Wellenfeldattribute, und damit ihre Brauchbarkeit für die tomographische Inversion, hängt von einer Reihe von Faktoren ab. Zu beachten ist insbesondere die Größe der während der Kohärenzanalyse des *CRS stack* verwendeten Apertur. Diese muss geeignet gewählt werden, um optimale kinematische Wellenfeldattribute zu erhalten. Bei starken lateralen Geschwindigkeitsvariationen werden Reflexionsereignisse möglicherweise nur noch schlecht durch Laufzeitapproximationen zweiter Ordnung beschrieben. Verlässliche Wellenfeldattribute können in solchen Fällen nicht bestimmt werden.

Um unphysikalische Fluktuationen in den kinematischen Wellenfeldattributen zu beseitigen ist es unter Umständen sinnvoll, die Attributsektionen ereigniskonsistent zu glätten. Eine solche Glättung wirkt sich sowohl auf das Stapelergebnis (Abbildung G.1), als auch auf die Anwendbarkeit der Attribute für die tomographische Inversion positiv aus. Beim Picken von Reflexionsereignissen müssen multiple Reflexionen nach Möglichkeit vermieden werden, da sie das Inversionsergebnis negativ beeinflussen. Die gepickten Datenpunkte sollten vor Anwendung der tomographischen Inversion editiert, das heißt, Ausreißer in den Daten, sowie Datenpunkte, die als Multiple identifiziert werden können (Abbildung 6.1), entfernt werden.

Der gesamte Ablauf der Bestimmung eines Geschwindigkeitsmodells mittels der beschriebenen tomographischen Inversion wird für den 2D Fall an einem synthetischen und einem realen seismischen Datenbeispiel demonstriert. Dazu wird zunächst ein mehrfach überdeckter synthetischer 2D seismischer Datensatz durch *ray tracing* in einem lateral inhomogenen Geschwindigkeitsmodell (Abbildung 6.2) erzeugt. Mithilfe der tomographischen Inversion soll dann aus den seismischen Daten ein konsistentes Geschwindigkeitsmodell konstruiert werden, das dem tatsächlichen Modell kinematisch möglichst äquivalent ist. Dazu werden die oben beschriebenen Bearbeitungsschritte (*CRS stack*, Picken) durchgeführt und die tomographische Inversion angewendet. Das resultierende Geschwindigkeitsmodell (Abbildungen 6.8a und b) ähnelt einer geglätteten Version der wahren Geschwindigkeitsverteilung. Die Lokationen der zu den verwendeten Datenpunkten gehörigen rekonstruierten NIP Punkte stimmen sehr gut mit den Geschwindigkeitsdiskontinuitäten im wahren Modell überein (Abbildung 6.8c). Das rekonstruierte Geschwindigkeitsmodell ist für die betrachteten Datenpunkte also kinematisch korrekt. Die gute Qualität des Inversionsergebnisses wird bestätigt durch die Anwendung einer Tiefenmigration auf die ungestapelten Daten. In einem konsistenten Modell sollte das Migrationsergebnis kinematisch unabhängig vom verwendeten *offset* sein. Eine Betrachtung des Migrationsergebnisses als Funktion des *offsets* an ausgewählten Lokationen (in sogenannten *common-image gathers*) zeigt, dass die Reflektortiefen in der Tat *offset*-unabhängig sind (Abbildung 6.10). Das gefundene Geschwindigkeitsmodell ist also konsistent mit den seismischen Daten.

Schließlich wird die tomographischen Inversion an einem realen seismischen Datenbeispiel getestet. Dazu wird ein von der Firma HotRock zur Verfügung gestellter, in der Nähe von Karlsruhe entlang eines 2D Profils gemessener seismischer Datensatz verwendet. Wiederum wird auf die vorprozessierten Daten zunächst der *CRS stack* (in diesem Fall gefolgt von einer Glättung der Attribute) angewendet (Abbildungen 6.12 und 6.13), eine Reihe von Punkten in der simulierten *zero-offset* Sektion gepickt und die zugehörigen kinematischen Wellenfeldattribute extrahiert. Mit den resultierenden (editierten) Datenpunkten (Abbildungen 6.14a, c und e) wird dann die tomographische Inversion durchgeführt. Ein Vergleich des Inversionsergebnisses (Abbildung 6.16) mit der tatsächlichen Geschwindigkeitsverteilung ist im Fall von Realdaten nicht möglich. Aussagen über die Qualität des Ergebnisses lassen sich daher nur mithilfe einer Tiefenmigration der ungestapelten Daten machen. Im vorliegenden Fall sind die Ereignisse in den betrachteten *common-image gathers* (Abbildung 6.18) weitgehend flach (*offset*-unabhängig). Das aus der tomographischen Inversion resultierende Geschwindigkeitsmodell ist also konsistent mit den seismischen Daten. Damit ist gezeigt, dass sich die vorgestellte tomographische Inversionsmethode auch unter realen Bedingungen erfolgreich zur Bestimmung eines Geschwindigkeitsmodells verwenden lässt.

## Schlussfolgerungen und Ausblick

In der vorliegenden Arbeit wurde eine neue tomographische Methode zur Bestimmung seismischer Geschwindigkeitsmodelle für die Tiefenmigration vorgestellt. Die Methode basiert auf der Verwendung von Laufzeitinformation in Form von kinematischen Wellenfeldattributen, die zum Beispiel mit der *CRS-stack* Methode aus den seismischen Daten bestimmt werden können. Damit lässt sich einer der Schwachpunkte konventioneller Reflexionstomographie, das Picken von Laufzeiten in den ungestapelten seismischen Daten, vermeiden. Stattdessen kann das Picken in einer gestapelten Sektion (der *CRS-stack* Sektion) mit erheblich verbessertem S/R Verhältnis durchgeführt werden. Die zugehörigen kinematischen Wellenfeldattribute werden dann automatisch aus den entsprechenden Attributsektionen extrahiert. Aufgrund der verwendeten Modellparametrisierung (glattes Geschwindigkeitsmodell und isolierte Reflexionspunkte bzw. NIP Lokationen) können die Picklokationen unabhängig voneinander betrachtet werden. Das Picken entlang kontinuierlichen Reflektoren ist nicht notwendig. Zu beachten ist, dass die Verwendung von kinematischen Wellenfeldattributen, und damit von Laufzeitapproximationen zweiter Ordnung, generell zu Einschränkungen in der Anwendbarkeit der Methode im Fall starker lateraler Geschwindigkeitsvariationen führt.

Sowohl die allgemeine Theorie als auch praktische Aspekte der tomographischen Inversion für den 1D, 2D und 3D Fall wurden diskutiert. Der gesamte Prozess der Modellbestimmung, ausgehend von den ungestapelten seismischen Daten, über die Anwendung des *CRS stack* und das Picken der Datenpunkte bis hin zur Tomographie selbst wurde an einem synthetischen und an einem realen seismischen Datenbeispiel demonstriert.

Zukünftige Arbeiten sollten sich auf praktische Aspekte der Anwendung der tomographischen Inversion auf reale 3D seismische Daten konzentrieren. Außerdem bleibt das Problem möglicher multipler Reflexionen in den Daten. Schließlich sollten Möglichkeiten einer weitergehenden Automatisierung des Inversionsprozesses untersucht werden.

# Contents

# Chapter 1

# Introduction

## 1.1 Seismic reflection imaging

The seismic reflection method plays an important role in the exploration for hydrocarbon resources in the Earth's interior. It is based on the concept of gaining information on subsurface geological structures and properties by bringing elastic energy (in the form of transient pulses) into the ground with controlled seismic sources and measuring the seismic wavefield that reaches the Earth's surface after reflection at discontinuities of the elastic properties in the subsurface. The recorded seismic wavefield allows to obtain detailed information on the spatial distribution of these discontinuities and, thus, on geological structures. In addition, an interpretation of the amplitudes of the measured wavefield may lead to quantitative estimates of the elastic parameters themselves. Both types of information are useful for identifying and delineating possible hydrocarbon reservoirs.

### 1.1.1 Seismic data acquisition

The exploration for oil and gas usually takes place in sedimentary rocks with target depths in the order of a few (up to 5) kilometers. Measurements are carried out on land as well as at sea. They normally involve a large number of seismic experiments, each consisting of a seismic source and an array of seismic receivers at the measurement surface with a range of different source-receiver separations (offsets), see Figure 1.1. Maximum offsets of several kilometers are usually used. In between experiments, the source or the entire source-receiver setup is moved (ideally a regular increment) to a new location. This leads to a seismic illumination of subsurface structures by several overlapping experiments and, thus, yields redundant information on these structures. The resulting seismic dataset is called multicoverage dataset.

In land seismic acquisition, sources may consist of explosives in shallow boreholes or of seismic vibrators bringing a frequency-modulated sweep signal into the subsurface, while the receivers (geophones) measure one or more components of particle motion (or some function thereof) due to the emerging elastic wavefield. Depending on whether 2D or 3D seismic measurements are carried out, the receivers are placed either along a single profile or in a two-dimensional array.

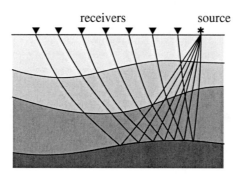

Figure 1.1: A seismic experiment involving one shot and an array of receivers. Only rays reflected from the lowermost interface are displayed.

In marine seismic acquisition, the source usually consists of an array of airguns, towed behind a vessel, which release high-pressure air pulses into the water. The emerging reflected acoustic wavefield is measured at a number of hydrophones, sensitive to pressure variations, which are placed in towed streamer cables of several kilometers length. In 2D marine acquisition, only one streamer is used, while in 3D marine acquisition, several streamers are towed behind the vessel in parallel. Apart from these conventional acquisition geometries, a range of other geometries, involving receivers in boreholes or placed at the sea floor, are possible.

### 1.1.2 Seismic data processing

The recorded seismic wavefield initially includes a whole range of different wave types. However, for seismic reflection imaging, only primary body wave reflection events (waves that have been reflected only once in the subsurface) of a specified wave mode are used. These are usually compressional waves, but may also be waves that have been converted at discontinuities on their way through the subsurface. All other wave types, including multiply reflected waves (multiples), surface waves, refracted waves, and primary reflections of other wave modes are treated as coherent noise. One of the aims of seismic reflection data processing is to suppress these unwanted waves and other types of coherent and random noise in the data and enhance the primary reflection signals to be used for producing a structural image of the subsurface by a process called migration.

A typical basic seismic processing sequence (e. g., Yilmaz, 1987) involves the following processing steps: preprocessing and filtering of the seismic data for noise removal, application of deconvolution to increase temporal resolution by removing the source signature, and to remove short period multiples and reverberations, sorting of the data into CMP gathers (collections of traces corresponding to a common midpoint between sources and receivers), stacking velocity analysis in the CMP gathers and stacking of signals in the CMP gathers to obtain a stacked section. The stacking velocity analysis and CMP stacking processes will be described in more detail in Section 3.2. They have the aim of suppressing random and coherent noise in the data and reducing the amount of data for further processing. In addition, the stacking velocities can be used as the basis for constructing a velocity model required for the next step in the processing sequence, the application of migration to obtain a structural image of the subsurface from the seismic data, as discussed below.

## 1.2 Depth imaging and the role of seismic velocity models

The main goal of seismic reflection imaging is to obtain detailed information on subsurface geological structures in the form of a seismic depth image. Such an image can be obtained from the preprocessed seismic data by a process called depth imaging or depth migration. This process is applied either to the stacked data, assuming them to represent a simulated zero-offset section or volume (poststack migration), or directly to the seismic data before stacking (prestack migration).

A range of different depth migration methods exist, all of which are based on the concept of using the wave equation to propagate or continue the recorded wavefield back into the subsurface and applying an imaging condition to obtain the depth image from the extrapolated wavefield. During this process, the effects of wave propagation through the subsurface, present in the recorded wavefield, are undone. Diffraction events in the seismic data are focused and reflectors are placed in their proper subsurface locations. Depth migration can, for example, be implemented based on an integral solution of the wave equation. It is then known as Kirchhoff migration (e. g., Schneider, 1978; Schleicher et al., 1993). Other migration approaches are based on the numerical downward continuation of the recorded wavefield in the space-time, frequency-space, or frequency-wavenumber domain (e. g., Gazdag, 1978; Claerbout, 1985; Stoffa et al., 1990).

However, because it is based on the extrapolation of wavefields, the application of depth migration to transform the measured seismic data into a structural depth image requires a model of the distribution of seismic wave velocities in the subsurface. Such a velocity model is usually initially unknown and needs to be constructed from the available information: possible geological or geophysical a priori knowledge of the investigated area, information from boreholes, and the seismic reflection data themselves. If an incorrect velocity model is used, the migrated image will not be properly focused and reflectors will not be placed in their correct spatial locations. While it is in principle not possible to derive the true subsurface velocity distribution from seismic data recorded on the Earth's surface, one can make use of the redundant information contained in seismic multi-coverage data to construct a velocity model that is consistent with these data. Velocity models thus obtained are also known as macro velocity models, as they are optimized for depth imaging and (although they should be geologically reasonable) are not directly suitable for detailed geological interpretation.

## 1.3 Velocity model estimation methods

A wide variety of different velocity model estimation methods are in use which are all ultimately based on the criterion of consistency with the seismic data, but differ in the way this consistency

3

is measured, in the model description, and in how deviations from consistency are translated into model updates. The different model parametrizations and model update relations correspond to different approximations and assumptions made about the subsurface. Not all methods are equally well suited for all geological environments and all levels of subsurface complexity. In addition, the different demands on data quality need to be considered.

The problem of determining velocity models from seismic reflection data is nonlinear, as the kinematics of wave propagation depend nonlinearly on the velocity distribution. Consequently, virtually all velocity model estimation techniques proceed iteratively, either by updating an initial model locally or globally, or by performing layer-stripping.

In the simplest case of horizontal reflectors in a medium with only vertical velocity variations (1D case), the stacking velocity values obtained from stacking velocity analysis in a CMP gather (Section 3.2) are sufficient to construct a 1D interval velocity model by applying Dix inversion (Dix, 1955). In laterally inhomogeneous media, more sophisticated methods need to be applied. Dix inversion is in such cases usually used to obtain a simple initial model which is then updated.

Methods for the construction of velocity models for depth imaging can be roughly divided into two major classes, based on how the criterion of consistency of the model with the seismic data is measured and used to obtain a model update: methods that make use of the focusing properties of prestack migration and methods that use traveltime information obtained from the seismic data. In the following, these two classes of velocity estimation methods will be briefly discussed. It should, however, be noted that this classification into migration-based methods and traveltime inversion is not strictly applicable, as many velocity estimation methods combine aspects of both approaches (e.g., Stork, 1992; Kosloff et al., 1996; Audebert et al., 1997) or use other ways of evaluating the consistency between the model and the seismic data (e.g., Landa et al., 1988; Jin and Madariaga, 1994). The references given in the following by no means represent a complete overview of available velocity model estimation techniques. Practical aspects of velocity model building are discussed in Fagin (1999) and Yilmaz (2001).

## 1.3.1   Migration-based velocity analysis

In velocity model estimation methods based on prestack migration the consistency of the model with the seismic data is evaluated by examining the result of prestack depth migration as a function of source-receiver offset. In a consistent model, seismic images obtained by prestack depth migration should be kinematically independent of the used offset (e.g., Gardner et al., 1974). Deviations from this consistency can be observed in the form of residual moveout in common-image gathers (CIGs), that is, in the offset- or angle-dependent migration result for fixed image locations, produced by common-shot migration, common-offset migration, or other migration types producing offset- or angle-dependent output gathers. If the velocity above a reflector in the model is too low, the corresponding event in a CIG curves upwards with increasing offset, while for velocities that are too high, events in a CIG curve downwards. In a consistent velocity model, reflection events in CIGs are flat, that is, independent of offset (Figure 1.2). Signals associated with such a flat event in a CIG can then be regarded as pertaining to one and the same reflection point in the subsurface. The moveout of events in CIGs obtained by prestack migration can be evaluated using the same

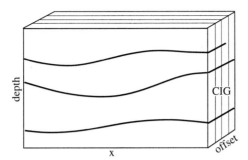

Figure 1.2: In migration-based velocity analysis, the prestack migration result is examined in common-image gathers (CIGs). In a consistent velocity model, the migration result shot be kinematically independent of offset. Consequently, events in CIGs should be flat.

tools that are used for stacking velocity analysis in CMP gathers (Section 3.2), namely analysis of semblance, equation (3.5), or other coherence measures along predefined moveout curves.

Al-Yahya (1989) and Deregowski (1990) propose simple model update procedures based on residual moveout analysis in CIGs. However, because of a number of simplifying assumptions made in these procedures (lateral velocity homogeneity, small offset, zero dip), they are only applicable for relatively simple models. Lafond and Levander (1993) and Liu (1997) present updating procedures which overcome these limitations by taking into account lateral velocity variations and reflector dips. Tomographic model update procedures based on residual moveouts in CIGs will be discussed below.

A different way of evaluating the results of prestack depth migration with respect to the velocity model has been proposed by Faye and Jeannot (1986), based on earlier work by Yilmaz and Chambers (1984), and has become known as depth focusing analysis (MacKay and Abma, 1992). It makes use of the fact that if migration is performed by alternate downward continuation of shots and receivers (shot-geophone migration) in the correct model, the energy related to a reflection event collapses (focuses) to zero offset at zero traveltime, which is the imaging condition used in the migration. If the velocity model is not correct, focusing occurs at a different depth. This depth deviation can be used (under the assumption of low dip and constant velocity) to calculate the true reflector depth and velocity value. Limitations due to the simplifying assumptions have been addressed by MacKay and Abma (1993) and Audebert and Diet (1993).

Both approaches of using migration for velocity model estimation require the repeated application of prestack migration, which is computationally expensive, especially in 3D.

## 1.3.2 Traveltime inversion/reflection tomography

Another way of applying the criterion of consistency with the seismic data for the estimation of velocity models is to consider the traveltimes of selected reflection events picked in the prestack data.

5

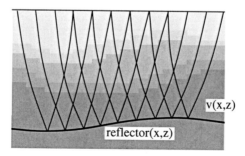

Figure 1.3: In classical reflection tomography, the gridded velocity model and reflectors are globally updated to minimize the misfit between traveltimes picked in the seismic data and those calculated along rays in the model.

A model, consisting of reflector interfaces and velocities, is then found iteratively by minimizing the misfit (usually in the least squares sense) between the picked traveltimes and the corresponding values calculated by ray tracing in the model.

Early traveltime inversion algorithms have been published by Sattlegger et al. (1981) and Gjøystdal and Ursin (1981). Different variants of classical reflection tomography (Figure 1.3) using smooth or gridded velocity models defined independently of the reflector geometry are described, among others, by Bishop et al. (1985), Farra and Madariaga (1988), Williamson (1990), and Stork and Clayton (1991). In these algorithms, the nonlinear optimization problem of minimizing the traveltime misfit is solved iteratively by computing global updates of the velocity field and reflector depths as solutions to large, linearized systems of equations. One of the drawbacks of reflection tomography is the large number of traveltime picks that is required. Picking usually needs to be performed in the prestack data along interpreted horizons corresponding to the reflectors in the model.

Modern reflection tomography implementations often combine tomographic traveltime inversion with migration-based velocity analysis (e. g., Stork, 1992; Woodward et al., 1998). For that purpose, the offset-dependent depth deviations (residual moveouts) in CIGs are converted into traveltime deviations using local reflector dip information. These traveltime deviations, related to the corresponding specular rays traced through the model are then used to calculate global velocity model updates by performing a tomographic inversion. Thus, the model is directly optimized to minimize the residual moveouts in CIGs. During each iteration, prestack depth migration needs to be performed and residual moveouts need to be picked in the considered CIGs.

A different approach to tomographic inversion called stereotomography has been introduced by Billette and Lambaré (1998), see also Billette et al. (2003). In stereotomography, local time dip (slope) information at the source and receiver side is used along with the traveltime for selected locally coherent reflection events in the prestack data. Associated with each such data point is a reflection or diffraction point in the subsurface, defined by its spatial location and reflection/diffraction angles which are treated as additional model parameters. During the inversion process, a model consisting of a smooth velocity distribution and parameters associated with the reflection/diffraction points (reflection point locations and reflection/diffraction angles) are globally iteratively updated to minimize the misfit between the observed and forward modeled values of reflection traveltime, time dip at the source and receiver points and the spatial locations of source and receiver points for each data point. Advantages of this method include the fact that, due to the additional dip information, only locally coherent events are required and no migrations are involved.

## 1.4   A new method for velocity model estimation

In practice, the velocity model estimation methods discussed in the previous section have a number of drawbacks. Migration-based methods require the repeated application of prestack depth migration, which is very expensive in terms of computation time, especially in 3D. After each iteration, the migration results need to be examined, usually interactively, in order to obtain information for updating the model. Methods based on tomographic inversion normally require kinematic information to be obtained by picking of events in the seismic prestack data, often along continuous, interpreted horizons. Particularly in the case of 3D seismic data, this picking process is very time-consuming due to the huge number of seismic traces involved. If the overall signal-to-noise (S/N) ratio is low, which is often the case in unstacked seismic data, the reliable identification of reflection events can become difficult or even impossible.

The main problem in practice is, thus, the efficient extraction of traveltime information for tomographic inversion from large seismic datasets with a possibly low S/N ratio. An effective way of addressing this problem is the use of second-order traveltime approximations. These provide a meaningful way of parameterizing reflection events and, thus, allow a stable determination of traveltimes, even in situations of low S/N ratio, by performing coherence analyses in the prestack data. In addition, the amount of data to be handled is significantly reduced, as reflection traveltimes can be represented by a small number of parameters, namely the coefficients of the traveltime approximations. Traveltime information in that form has been used for the determination of seismic velocities for a long time in the form of Dix inversion based on stacking velocities (Dix, 1955) and related methods (e. g., Hubral and Krey, 1980). Although there are limitations to the applicability of second-order traveltime approximations in cases of strong lateral velocity variations, such approximations have proven to be very useful also in laterally inhomogeneous media and may, in fact, in certain situations be the only way of obtaining reliable kinematic information for velocity model estimation.

Second-order traveltime information useful for the determination of laterally inhomogeneous velocity models may for example be extracted from the seismic data with the common-reflection-surface (CRS) stack technique (e. g., Mann et al., 1999; Jäger et al., 2001), which can be regarded as a generalization of conventional stacking velocity analysis. The CRS stack is based on the use of stacking surfaces in the form of second-order traveltime approximations in the midpoint and half-offset coordinates to obtain a simulated zero-offset section (2D case) or volume (3D case). For each zero-offset sample, a set of parameters, called kinematic wavefield attributes, describing the optimum stacking surface is determined automatically with a coherence analysis in the multi-coverage data. These kinematic wavefield attributes can be interpreted in terms of the second-order traveltimes of two hypothetical emerging wavefronts at the respective zero-offset location. One of these, the so-called NIP wavefront, is due to a hypothetical point source at the normal-incidence point (NIP) of the associated zero-offset ray. This wave is very useful for the determination of velocity models as it provides a clear criterion for their consistency.

In this thesis, a new inversion method will be introduced which extends the concept of velocity model estimation based on second-order traveltime approximations to the tomographic determination of smooth, laterally inhomogeneous, isotropic velocity models in 2D and 3D. The method makes use of traveltime information in the form of the kinematic wavefield attributes related to hypothetical NIP waves obtained from the seismic prestack data with the CRS stack. In order to extract the required attributes from the CRS stack results, it is sufficient to perform picking of reflection events in the stacked, simulated zero-offset section/volume, in which events are much easier to identify than in the prestack data. Because of the special model parametrization used in the tomographic inversion, which is similar to that of stereotomography (Billette and Lambaré, 1998), pick locations in the simulated zero-offset section/volume can be considered independently of each other. They do not need to follow interpreted horizons, but may be located on reflection events that are only locally coherent. A relatively small number of picks are sufficient to perform the tomographic inversion. This significantly simplifies and speeds up the picking process and allows to obtain a velocity model even in situations in which it is not possible, due to a low S/N ratio or complex reflector structure, to identify reflection events continuously across the seismic section/volume.

## 1.5 Structure of the thesis

In this thesis, the complete theory for 1D, 2D, and 3D tomographic inversion with kinematic wavefield attributes is presented, practical and numerical aspects of the method are discussed, and its application is demonstrated on synthetic and real seismic data examples.

In **Chapter 1** (this chapter), a brief introduction to the basic concepts of reflection seismics is given. The importance of velocity models for seismic depth imaging is discussed and commonly used methods for the estimation of velocity models are reviewed, before proposing a new tomographic velocity model estimation method based on kinematic wavefield attributes.

**Chapter 2** gives an introduction to seismic ray theory with special emphasis on aspects that are relevant for the development of the tomographic inversion in subsequent chapters. These aspects include paraxial ray theory and ray perturbation theory, which are formulated for arbitrary curvilinear orthogonal coordinates. Paraxial ray tracing is discussed in more detail in two coordinate systems that are used in the tomographic inversion: the ray-centered coordinate system and the global Cartesian coordinate system.

In **Chapter 3**, the common-reflection-surface (CRS) stack method is introduced. Starting from the well-known CMP stack, the basic concepts of the CRS stack are discussed. The parameters of the traveltime approximation used in the CRS stack, the kinematic wavefield attributes, are interpreted in terms of two hypothetical emerging wavefronts. Finally, practical aspects of the application of the CRS stack are discussed.

In **Chapter 4**, the concept of tomographic inversion with kinematic wavefield attributes is introduced for the general 3D case. For that purpose, data and model components to be used in the inversion are defined, and the inverse problem is formulated as a nonlinear least-squares problem that is solved iteratively by local linearization during each iteration. The regularization of the inverse problem and the application of additional constraints are discussed. The inversion scheme is then formulated into an algorithm that can be used as the basis for implementing the method.

Implementational aspects of the tomographic inversion are treated in more detail in **Chapter 5** for the 1D case, the 2D case, and the 3D case with limited azimuth information. For each of these cases, the definition of data and model components, the forward modeling and calculation of Fréchet derivatives, the structure of the tomographic matrix and the solution of the resulting matrix equations are discussed. The inversion algorithm is demonstrated for each case by applying it to synthetic test data.

In **Chapter 6**, practical aspects of velocity model estimation with the tomographic inversion introduced in previous chapters are discussed. The entire inversion process, including the application of the CRS stack, the picking of input data, and the tomographic inversion itself, is then demonstrated on a synthetic and a real 2D seismic dataset. In both cases, a velocity model that is consistent with the data is found, which is verified by the application of prestack depth migration.

Finally, **Chapter 7** contains the conclusions and an outlook on future research concerning the presented tomographic inversion method. This includes further tests and applications to real 3D seismic data, the handling of picks related to multiples, and efforts towards further automatizing the entire inversion process.

Various topics, namely the physical interpretation of the kinematic wavefield attributes, the velocity model description in terms of B-splines, the regularization of the inverse problem, the calculation of Fréchet derivatives for the tomographic matrix, and a simple algorithm for smoothing kinematic wavefield attribute sections, are treated in **Appendices A** to **G**.

# Chapter 2

# Ray theory

In this chapter, a brief introduction to seismic ray theory as a tool for the description of high-frequency seismic wave propagation will be given. Emphasis will be put on results which are relevant to the tomographic inversion method to be introduced in Chapter 4. Of particular importance in this context is paraxial ray theory (Section 2.3), with which kinematic properties of waves calculated along a given ray may be approximately extrapolated into the vicinity of that ray. Also of great relevance for the tomographic inversion are the results of ray perturbation theory (Section 2.4). These allow to examine the first-order effects of perturbations of the velocity and other model parameters on the kinematics of wave propagation along a given ray. Paraxial ray theory and ray perturbation theory will be formulated in a form valid for arbitrary curvilinear orthogonal coordinate systems.

Although the presence of discontinuities of the elastic properties of the medium (piecewise smooth media) can in principle be handled by ray theory (e. g., Červený, 2001), a smoothly varying isotropic elastic medium without discontinuities will be assumed throughout this chapter. For paraxial ray tracing, continuous second spatial derivatives of velocity will be required, while for certain applications of ray perturbation theory, continuous third spatial derivatives are necessary.

Comprehensive treatments of a range of different aspects of ray theory can for example be found in Kravtsov and Orlov (1990) and Červený (2001).

## 2.1  High-frequency solutions of the elastodynamic wave equation

In seismology, it is usually assumed that for small-amplitude displacements the Earth can be described in terms of continuum mechanics as an elastic medium. Wave propagation in such a medium is governed by the elastodynamic equation (e. g., Aki and Richards, 1980). In the general anisotropic case, the elastodynamic equation, or wave equation, for the displacement vector component $u_i$ in the absence of body forces reads

$$\left(C_{ijkl}\, u_{k,l}\right)_{,j} = \rho\, \ddot{u}_i\,, \tag{2.1}$$

where $C_{ijkl}$ is the elastic tensor, containing, in the most general case, 21 independent parameters and $\rho$ is the density. Both, the elastic tensor and the density are functions of position, $C_{ijkl} = C_{ijkl}(x_m)$ and $\rho = \rho(x_m)$. Here, and throughout this section, subscripts denote Cartesian components and range from 1 to 3. In addition, the Einstein summation convention is used, that is, summation is carried out over repeated indices. The notation $u_{i,j}$ denotes the partial derivative $\partial u_i / \partial x_j$, and overdots are used to indicate time derivatives, as in $\ddot{u}_i = \partial^2 u_i / \partial t^2$. In the special case of an isotropic medium considered here, the elastic tensor takes the form

$$C_{ijkl} = \lambda \delta_{ij}\delta_{kl} + \mu(\delta_{ik}\delta_{jl} + \delta_{il}\delta_{jk}) \tag{2.2}$$

and only two independent parameters, the Lamé parameters $\lambda$ and $\mu$, remain ($\delta_{ij}$ is the Kronecker symbol). Inserting this expression into equation (2.1) yields

$$(\lambda + \mu)u_{j,ij} + \mu\, u_{i,jj} + \lambda_{,i}u_{j,j} + \mu_{,j}(u_{i,j} + u_{j,i}) = \rho\,\ddot{u}_i \,. \tag{2.3}$$

In inhomogeneous media, the wavefield cannot generally be separated into independently traveling waves and equation (2.3) is difficult to solve. To make the problem of describing and interpreting wavefields measured in reflection seismology tractable, the assumption of high-frequency wave propagation is often made, which allows to construct approximate solutions to the wave equation. High frequency in this context means that the dominant signal wavelengths are assumed to be small compared to characteristic length scales of heterogeneities of the medium. This assumption is justified by the fact that reflection events with well-defined traveltimes may be observed in recorded seismic data. As will be shown below, in the high-frequency limit, compressional and shear waves may propagate independently of each other, as in the case of a homogeneous medium, which is in accordance with observations.

To find an approximate high-frequency solution to equation (2.3), a time-harmonic trial solution of the form

$$u_i(x_j,t) = U_i(x_j)\exp\left[-i\omega\left(t - \tau(x_j)\right)\right] \tag{2.4}$$

is used. Here, $U_i(x_j)$ and $\tau(x_j)$ are smooth functions of $x_j$, and the circular frequency $\omega$ is assumed to be high. Alternatively, a trial solution in the form of a time-domain signal $u_i(x_j,t) = U_i(x_j)F\left(t - \tau(x_j)\right)$ may be used, where the Fourier spectrum of the analytical signal $F$ (e. g., Červený, 2001) is assumed to effectively vanish for small frequencies $\omega$. Substituting (2.4) for $u_i$ in equation (2.3) yields

$$(i\omega)^2 N_i(U_j) + i\omega M_i(U_j) + L_i(U_j) = 0 \tag{2.5}$$

with

$$\begin{aligned}
N_i(U_j) &= (\lambda + \mu)U_j\tau_{,i}\tau_{,j} + \mu U_i\tau_{,j}\tau_{,j} - \rho U_i \,, \\
M_i(U_j) &= (\lambda + \mu)\left[U_{j,i}\tau_{,j} + U_{j,j}\tau_{,i} + U_j\tau_{,ij}\right] \\
&\quad + \mu\left[2U_{i,j}\tau_{,j} + U_i\tau_{,jj}\right] + \lambda_{,i}U_j\tau_{,j} + \mu_{,j}(U_i\tau_{,j} + U_j\tau_{,i}) \,, \\
L_i(U_j) &= (\lambda + \mu)U_{j,ij} + \mu U_{i,jj} + \lambda_{,i}U_{j,j} + \mu_{,j}(U_{i,j} + U_{j,i}) \,.
\end{aligned} \tag{2.6}$$

For equation (2.5) to be satisfied for arbitrary high frequencies, each term must vanish independently. For high frequencies $\omega$, the first and second term dominate over the third term, which is, therefore, usually neglected. To take the third term in equation (2.5) into account more rigorously, a trial solution in the form of an asymptotic series in negative powers of $(i\omega)$ may be used instead of expression (2.4). Equation (2.4) then represents the zero-order term of the asymptotic series, leading to zero-order ray theory.

Setting $N_i(U_j)$ in equation (2.6) equal to zero and defining the slowness vector $p_i := \tau_{,i}$ yields an equation of the form

$$\left(\Gamma_{ij} - \delta_{ij}\right) U_j = 0 \quad \text{with} \quad \Gamma_{ij} = \left[\frac{(\lambda + \mu)}{\rho} p_i p_j + \frac{\mu}{\rho} p_k p_k \delta_{ij}\right] . \tag{2.7}$$

Equation (2.7) represents an eigenvalue problem for the matrix $\Gamma_{ij}$. It has non-trivial solutions if

$$\det\left(\Gamma_{ij} - \delta_{ij}\right) = \left(\frac{\mu}{\rho} p_k p_k - 1\right)^2 \left(\frac{(\lambda + 2\mu)}{\rho} p_k p_k - 1\right) = 0 , \tag{2.8}$$

which leads to the possible solutions

$$p_k p_k = \frac{\rho}{\mu} \qquad = v_s^{-2} , \tag{2.9a}$$

$$p_k p_k = \frac{\rho}{\lambda + 2\mu} = v_p^{-2} . \tag{2.9b}$$

Equations (2.9a) and (2.9b) are called *eikonal equations*. They describe the propagation of surfaces of constant $\tau$, or discontinuities of the wavefield, which may be interpreted as wavefronts, in an inhomogeneous medium. Obviously, in the high-frequency limit wavefronts associated with two different wave types may propagate independently. Equation (2.9a) describes the propagation of high-frequency shear waves (or S waves) with an S-wave velocity $v_s$, while (2.9b) describes the propagation of high-frequency compressional waves (P waves) with a P-wave velocity $v_p$. The corresponding mutually perpendicular eigenvectors give the polarization directions associated with these waves. The amplitude of $U_i$ can be obtained from setting $M_i(U_j) = 0$, which leads to the transport equations for P and S waves. These will, however, not be needed in the following sections, where only kinematic aspects of ray theory will be discussed, and are, therefore, not written out explicitly.

Alternatively to the described procedure, the eikonal equation may also be derived from Fermat's principle (e. g., Červený, 2001). However, in that case, there is initially no justification for treating P waves and S waves separately.

The conditions of validity of the high-frequency solution of the wave equation given in this section and of the ray theory results presented in the following sections are difficult to quantify. In general, it can be said that for ray theory to be valid, the involved signal wavelengths should be much smaller than the length scale of medium heterogeneities. A number of conditions for the validity of ray theory have been given by Ben-Menahem and Beydoun (1985a,b).

## 2.2 Solution of the eikonal equation

The eikonal equation derived for the isotropic case in the previous section is a first-order nonlinear partial differential equation. As noted above, it describes the propagation of discontinuities of the wavefield, which can be interpreted as wavefronts. Depending on the velocity used in the eikonal equation ($v_p$ or $v_s$), these wavefronts correspond to propagating P waves or S waves. All results of ray theory presented in this chapter are equally valid for both of these wave types. Therefore, the quantity $v$ used in the following may denote either the P wave or the S wave velocity. A common way of solving the eikonal equation is the method of characteristics (e. g., Courant and Hilbert, 1968; Bleistein, 1984).

### 2.2.1 Cartesian coordinates

If the eikonal equation in Cartesian coordinates $\left(x_1, x_2, x_3\right)$ is rewritten as

$$\tilde{H}(\mathbf{x}, \mathbf{p}^{(x)}) = \frac{v(\mathbf{x})}{2} \left( (\mathbf{p}^{(x)})^2 - \frac{1}{v^2(\mathbf{x})} \right) = 0 \qquad (2.10)$$

with $\mathbf{p}^{(x)} = \nabla \tau$, the method of characteristics leads to a system of six first-order ordinary differential equations, known as the characteristic equations or the Hamiltonian system, which are equivalent to the eikonal equation:

$$\frac{dx_i}{ds} = \frac{\partial \tilde{H}}{\partial p_i^{(x)}} = v p_i^{(x)} \qquad i = 1, 2, 3 ,$$

$$\frac{dp_i^{(x)}}{ds} = -\frac{\partial \tilde{H}}{\partial x_i} = -\frac{1}{v^2} \frac{\partial v}{\partial x_i} \qquad i = 1, 2, 3 . \qquad (2.11)$$

These equations describe curves—known as characteristics—in physical space, along which the differential equation (2.10) is satisfied, given appropriate initial conditions $\mathbf{x} = \mathbf{x}_0$ and $\mathbf{p}^{(x)} = \mathbf{p}_0^{(x)}$. In physical terms, the characteristics of the eikonal equation are rays, therefore equations (2.11) are also known as the ray-tracing system. The parameter $s$ is the running parameter along the rays and the function $\tilde{H}$ is called the Hamiltonian. An additional equation,

$$\frac{d\tau}{ds} = \sum_{i=1}^{3} p_i^{(x)} \frac{\partial \tilde{H}}{\partial p_i^{(x)}} = \frac{1}{v} , \qquad (2.12)$$

gives the traveltime along a ray. From equation (2.12) it can be seen that in the present case, $s$ has the physical meaning of arclength along the ray. Other choices of the Hamiltonian $\tilde{H}$ are possible, leading to different forms of the ray-tracing system with a different running parameter along the ray (e. g., Červený, 2001).

### 2.2.2 Curvilinear orthogonal coordinates

In general curvilinear orthogonal coordinates $(\zeta_1, \zeta_2, \zeta_3)$ with an infinitesimal line element

$$dS^2 = d\mathbf{r} \cdot d\mathbf{r} = h_1^2 d\zeta_1^2 + h_2^2 d\zeta_2^2 + h_3^2 d\zeta_3^2 \,, \tag{2.13}$$

the slowness vector is given by

$$\nabla\tau = \left( h_1^{-1}\frac{\partial\tau}{\partial\zeta_1}, h_2^{-1}\frac{\partial\tau}{\partial\zeta_2}, h_3^{-1}\frac{\partial\tau}{\partial\zeta_3} \right)^T \,, \tag{2.14}$$

where the scale factors $h_i$ are defined by $h_i = |\partial\mathbf{r}/\partial\zeta_i|$. The superscript $T$ denotes the transpose. With $p_i^{(\zeta)} := \partial\tau/\partial\zeta_i$, the eikonal equation then reads

$$\sum_{i=1}^{3} h_i^{-2}(p_i^{(\zeta)})^2 = \frac{1}{v^2(\zeta_1, \zeta_2, \zeta_3)} \,. \tag{2.15}$$

Note that in the general case, the $p_i^{(\zeta)}$ are not identical to the components of the slowness vector (2.14). The quantities $\zeta_i$, $p_i^{(\zeta)}$, $i = 1,2,3$ may be regarded as generalized coordinates in a six-dimensional phase space. If the eikonal equation is written as $\tilde{H}\left(\zeta_1, \zeta_2, \zeta_3, p_1^{(\zeta)}, p_2^{(\zeta)}, p_3^{(\zeta)}\right) = 0$, the ray-tracing system in curvilinear orthogonal coordinates is given by

$$\frac{d\zeta_i}{d\sigma} = \frac{\partial\tilde{H}}{\partial p_i^{(\zeta)}} \qquad i = 1,2,3 \,,$$

$$\frac{dp_i^{(\zeta)}}{d\sigma} = -\frac{\partial\tilde{H}}{\partial\zeta_i} \qquad i = 1,2,3 \,, \tag{2.16}$$

where $\sigma$ denotes the running parameter along the ray. The traveltime along a given ray can be obtained by integrating

$$\frac{d\tau}{d\sigma} = \sum_{i=1}^{3} p_i^{(\zeta)}\frac{\partial\tilde{H}}{\partial p_i^{(\zeta)}} \,. \tag{2.17}$$

The number of equations in the ray-tracing system can in general be reduced from six to four by using the eikonal equation to eliminate one of the space variables (e. g., Červený, 2001). By solving the eikonal equation (2.15) for $p_3^{(\zeta)}$, one obtains

$$p_3^{(\zeta)} = h_3\sqrt{\frac{1}{v^2} - h_1^{-2}(p_1^{(\zeta)})^2 - h_2^{-2}(p_2^{(\zeta)})^2} = -H\left(\zeta_1, \zeta_2, \zeta_3, p_1^{(\zeta)}, p_2^{(\zeta)}\right) \,. \tag{2.18}$$

The function $H$ will be called the reduced Hamiltonian. Defining

$$\tilde{H} = p_3^{(\zeta)} + H = 0 \tag{2.19}$$

and applying equations (2.16), leads to

$$\frac{d\zeta_i}{d\sigma} = \frac{\partial H}{\partial p_i^{(\zeta)}} \qquad i = 1,2 \,,$$

$$\frac{dp_i^{(\zeta)}}{d\sigma} = -\frac{\partial H}{\partial\zeta_i} \qquad i = 1,2 \,, \tag{2.20}$$

where, because of $\partial\zeta_3/\partial\sigma = 1$, the coordinate $\zeta_3$ can be used as the independent parameter $\sigma$ along the ray, and $p_3^{(\zeta)}$ can be directly calculated from (2.18), $p_3^{(\zeta)} = -H$. In the following, $\zeta_3 \equiv \sigma$ will be assumed. The reduced ray-tracing system (2.20) can be used, if $\zeta_3$ varies strictly monotonously along the ray, that is, if the ray has no turning point with respect to the $\zeta_3$ direction. The traveltime along the ray can then be obtained from

$$\frac{d\tau}{d\sigma} = \sum_{i=1}^{3} p_i^{(\zeta)} \frac{\partial\tilde{H}}{\partial p_i^{(\zeta)}} = \frac{h_3^2}{p_3^{(\zeta)}v^2} \, . \tag{2.21}$$

## 2.3   Paraxial ray tracing

Once the trajectory $\zeta_i(\sigma)$, $p_i^{(\zeta)}(\sigma)$, $i = 1,2$ of a ray (the central ray) has been determined, the behavior of rays with coordinates

$$\begin{aligned} \zeta_i + \Delta\zeta_i \qquad & i = 1,2 \\ p_i^{(\zeta)} + \Delta p_i^{(\zeta)} \qquad & i = 1,2 \end{aligned} \tag{2.22}$$

in its close vicinity can be described by a second-order approximation of the eikonal equation around the central ray. Inserting (2.22) into the Hamiltonian system (2.20) and expanding the right-hand side of (2.20) into a Taylor series up to first order leads to the linear system of equations

$$\begin{aligned} \frac{d\Delta\zeta_i}{d\sigma} &= \sum_{j=1}^{2} \left( \frac{\partial^2 H}{\partial p_i^{(\zeta)}\partial\zeta_j}\Delta\zeta_j + \frac{\partial^2 H}{\partial p_i^{(\zeta)}\partial p_j^{(\zeta)}}\Delta p_j^{(\zeta)} \right) \quad i = 1,2\,, \\ \frac{d\Delta p_i^{(\zeta)}}{d\sigma} &= -\sum_{j=1}^{2} \left( \frac{\partial^2 H}{\partial\zeta_i\partial\zeta_j}\Delta\zeta_j + \frac{\partial^2 H}{\partial\zeta_i\partial p_j^{(\zeta)}}\Delta p_j^{(\zeta)} \right) \quad i = 1,2\,, \end{aligned} \tag{2.23}$$

which is known as the paraxial ray-tracing system. Note that all partial derivatives of $H$ are evaluated on the central ray. If a vector $\Delta\boldsymbol{\eta} = \left(\Delta\zeta_1, \Delta\zeta_2, \Delta p_1^{(\zeta)}, \Delta p_2^{(\zeta)}\right)^T$ and a matrix $\underline{\mathbf{S}}$ with

$$\underline{\mathbf{S}} = \begin{pmatrix} \dfrac{\partial^2 H}{\partial p_1^{(\zeta)}\partial\zeta_1} & \dfrac{\partial^2 H}{\partial p_1^{(\zeta)}\partial\zeta_2} & \dfrac{\partial^2 H}{\partial p_1^{(\zeta)}\partial p_1^{(\zeta)}} & \dfrac{\partial^2 H}{\partial p_1^{(\zeta)}\partial p_2^{(\zeta)}} \\[2ex] \dfrac{\partial^2 H}{\partial p_2^{(\zeta)}\partial\zeta_1} & \dfrac{\partial^2 H}{\partial p_2^{(\zeta)}\partial\zeta_2} & \dfrac{\partial^2 H}{\partial p_2^{(\zeta)}\partial p_1^{(\zeta)}} & \dfrac{\partial^2 H}{\partial p_2^{(\zeta)}\partial p_2^{(\zeta)}} \\[2ex] -\dfrac{\partial^2 H}{\partial\zeta_1\partial\zeta_1} & -\dfrac{\partial^2 H}{\partial\zeta_1\partial\zeta_2} & -\dfrac{\partial^2 H}{\partial\zeta_1\partial p_1^{(\zeta)}} & -\dfrac{\partial^2 H}{\partial\zeta_1\partial p_2^{(\zeta)}} \\[2ex] -\dfrac{\partial^2 H}{\partial\zeta_2\partial\zeta_1} & -\dfrac{\partial^2 H}{\partial\zeta_2\partial\zeta_2} & -\dfrac{\partial^2 H}{\partial\zeta_2\partial p_1^{(\zeta)}} & -\dfrac{\partial^2 H}{\partial\zeta_2\partial p_2^{(\zeta)}} \end{pmatrix} \tag{2.24}$$

are defined, the linear system of equations (2.23) can be written as

$$\frac{d\Delta\boldsymbol{\eta}}{d\sigma} = \underline{\mathbf{S}}\Delta\boldsymbol{\eta} \, . \tag{2.25}$$

The range of validity of the paraxial ray approximation away from the central ray depends on how well the Hamiltonian $H$ is approximated around the central ray by its second-order Taylor expansion. The quality of this approximation is mainly controlled by the degree of inhomogeneity of the velocity distribution near the central ray.

### 2.3.1 Ray propagator matrix

Due to the linearity of the system (2.25), its general solution may be written in terms of a fundamental matrix. Thus, if a $4 \times 4$ matrix $\underline{\mathbf{\Pi}}^{(\zeta)}(\sigma, \sigma_0)$ with $\underline{\mathbf{\Pi}}^{(\zeta)}(\sigma_0, \sigma_0) = \underline{\mathbf{I}}_4$ (the $4 \times 4$ identity matrix) is introduced, which solves

$$\frac{d}{d\sigma} \underline{\mathbf{\Pi}}^{(\zeta)} = \underline{\mathbf{S}} \, \underline{\mathbf{\Pi}}^{(\zeta)} \,, \tag{2.26}$$

the solution $\Delta\boldsymbol{\eta}(\sigma)$ of equation (2.25) for any initial conditions $\Delta\boldsymbol{\eta}(\sigma_0)$ may be expressed as

$$\Delta\boldsymbol{\eta}(\sigma) = \underline{\mathbf{\Pi}}^{(\zeta)}(\sigma, \sigma_0) \Delta\boldsymbol{\eta}(\sigma_0) \,. \tag{2.27}$$

The matrix $\underline{\mathbf{\Pi}}^{(\zeta)}$ is also called ray propagator matrix (e. g., Červený, 2001).

The ray propagator matrix has a number of properties that will be briefly discussed in the following. Details can be found in Červený (2001). To start with, the $4 \times 4$ matrix $\underline{\mathbf{\Pi}}^{(\zeta)}$ will be written in terms of four $2 \times 2$ submatrices:

$$\underline{\mathbf{\Pi}}^{(\zeta)} = \begin{pmatrix} \underline{\mathbf{Q}}_1^{(\zeta)} & \underline{\mathbf{Q}}_2^{(\zeta)} \\ \underline{\mathbf{P}}_1^{(\zeta)} & \underline{\mathbf{P}}_2^{(\zeta)} \end{pmatrix} \,. \tag{2.28}$$

Due to the structure of $\underline{\mathbf{S}}$, the matrix $\underline{\mathbf{\Pi}}^{(\zeta)}$ can be shown to possess the following so-called *symplectic property*:

$$\underline{\mathbf{\Pi}}^{(\zeta)T} \underline{\mathbf{J}} \, \underline{\mathbf{\Pi}}^{(\zeta)} = \underline{\mathbf{J}} \,, \quad \text{where} \quad \underline{\mathbf{J}} = \begin{pmatrix} \mathbf{0}_2 & \mathbf{I}_2 \\ \mathbf{I}_2 & \mathbf{0}_2 \end{pmatrix} \,. \tag{2.29}$$

Here, $\underline{\mathbf{I}}_2$ is the $2 \times 2$ unit matrix and $\underline{\mathbf{0}}_2$ is the $2 \times 2$ zero matrix. This leads to a number of invariants that remain constant along the ray:

$$\begin{aligned}
\underline{\mathbf{Q}}_1^{(\zeta)T} \underline{\mathbf{P}}_1^{(\zeta)} - \underline{\mathbf{P}}_1^{(\zeta)T} \underline{\mathbf{Q}}_1^{(\zeta)} &= \mathbf{0}_2 \,, & \underline{\mathbf{P}}_2^{(\zeta)T} \underline{\mathbf{Q}}_1^{(\zeta)} - \underline{\mathbf{Q}}_2^{(\zeta)T} \underline{\mathbf{P}}_1^{(\zeta)} &= \mathbf{I}_2 \,, \\
\underline{\mathbf{Q}}_2^{(\zeta)T} \underline{\mathbf{P}}_2^{(\zeta)} - \underline{\mathbf{P}}_2^{(\zeta)T} \underline{\mathbf{Q}}_2^{(\zeta)} &= \mathbf{0}_2 \,, & \underline{\mathbf{Q}}_1^{(\zeta)T} \underline{\mathbf{P}}_2^{(\zeta)} - \underline{\mathbf{P}}_1^{(\zeta)T} \underline{\mathbf{Q}}_2^{(\zeta)} &= \mathbf{I}_2 \,.
\end{aligned} \tag{2.30}$$

From the symplectic property it follows that the *inverse* of $\underline{\mathbf{\Pi}}^{(\zeta)}(\sigma, \sigma_0)$ can be written in terms of $\underline{\mathbf{Q}}_1^{(\zeta)}, \underline{\mathbf{Q}}_2^{(\zeta)}, \underline{\mathbf{P}}_1^{(\zeta)}$, and $\underline{\mathbf{P}}_2^{(\zeta)}$ as

$$\underline{\mathbf{\Pi}}^{(\zeta)-1}(\sigma, \sigma_0) = \underline{\mathbf{\Pi}}^{(\zeta)}(\sigma_0, \sigma) = \begin{pmatrix} \underline{\mathbf{P}}_2^{(\zeta)T}(\sigma, \sigma_0) & -\underline{\mathbf{Q}}_2^{(\zeta)T}(\sigma, \sigma_0) \\ -\underline{\mathbf{P}}_1^{(\zeta)T}(\sigma, \sigma_0) & \underline{\mathbf{Q}}_1^{(\zeta)T}(\sigma, \sigma_0) \end{pmatrix} \,. \tag{2.31}$$

It can also be shown (e. g., Gilbert and Backus, 1966) that the ray propagator matrix satisfies the *chain rule*

$$\underline{\mathbf{\Pi}}^{(\zeta)}(\sigma, \sigma_0) = \underline{\mathbf{\Pi}}^{(\zeta)}(\sigma, \sigma_1) \, \underline{\mathbf{\Pi}}^{(\zeta)}(\sigma_1, \sigma_0) \,, \tag{2.32}$$

where $\sigma_1$ does not need to lie between $\sigma$ and $\sigma_0$.

### 2.3.2 Dynamic ray tracing

In 3D media, the system of rays corresponding to a certain specified wave may be parameterized by two parameters $\gamma_1$ and $\gamma_2$. In the case of a point source, these parameters may for example be the angles of the spherical polar coordinate system. The purpose of dynamic ray tracing is to determine the partial derivatives of the coordinates $\zeta_i$, $p_i^{(\zeta)}$, $i = 1, 2$ with respect to these parameters $\gamma_1$ and $\gamma_2$ or any other initial parameters of a ray. If the $2 \times 2$ matrices $\underline{\mathbf{Q}}^{(\zeta)}$ and $\underline{\mathbf{P}}^{(\zeta)}$ with

$$Q_{ij}^{(\zeta)} = \frac{\partial \zeta_i}{\partial \gamma_j}, \qquad P_{ij}^{(\zeta)} = \frac{\partial p_i^{(\zeta)}}{\partial \gamma_j} \qquad i = 1, 2, \quad j = 1, 2 \tag{2.33}$$

are introduced, it follows from equations (2.23) that these matrices satisfy the system of equations

$$\frac{d}{d\sigma} \begin{pmatrix} \underline{\mathbf{Q}}^{(\zeta)} \\ \underline{\mathbf{P}}^{(\zeta)} \end{pmatrix} = \underline{\mathbf{S}} \begin{pmatrix} \underline{\mathbf{Q}}^{(\zeta)} \\ \underline{\mathbf{P}}^{(\zeta)} \end{pmatrix}, \tag{2.34}$$

known as the dynamic ray-tracing system (e. g. Červený, 2001), which is identical in form to the paraxial ray-tracing system (2.25). The submatrix $\left( \underline{\mathbf{Q}}_1^{(\zeta)}, \underline{\mathbf{P}}_1^{(\zeta)} \right)^T$ of the ray propagator matrix $\underline{\boldsymbol{\Pi}}^{(\zeta)}$ is, thus, a solution of the system (2.34) for initial conditions $\left( \mathbf{I}_2, \mathbf{0}_2 \right)^T$, while $\left( \underline{\mathbf{Q}}_2^{(\zeta)}, \underline{\mathbf{P}}_2^{(\zeta)} \right)^T$ is a solution for initial conditions $\left( \mathbf{0}_2, \mathbf{I}_2 \right)^T$. The dynamic ray-tracing system has many applications, particularly in the computation of geometrical spreading, that is, the solution of the transport equation.

Using matrices $\underline{\mathbf{Q}}^{(\zeta)}$ and $\underline{\mathbf{P}}^{(\zeta)}$, the $2 \times 2$ matrix $\underline{\mathbf{M}}^{(\zeta)}$ of second derivatives of traveltime with respect to the coordinates $\zeta_1$ and $\zeta_2$ can be determined. The element $P_{ij}^{(\zeta)}$ of matrix $\underline{\mathbf{P}}^{(\zeta)}$ may be written as

$$P_{ij}^{(\zeta)} = \frac{\partial p_i^{(\zeta)}}{\partial \gamma_j} = \sum_{k=1}^{2} \frac{\partial p_i^{(\zeta)}}{\partial \zeta_k} \frac{\partial \zeta_k}{\partial \gamma_j} = \sum_{k=1}^{2} \frac{\partial^2 \tau}{\partial \zeta_i \partial \zeta_k} \frac{\partial \zeta_k}{\partial \gamma_j} = \sum_{k=1}^{2} M_{ik}^{(\zeta)} Q_{kj}^{(\zeta)} \quad i = 1, 2, \quad j = 1, 2. \tag{2.35}$$

Therefore, $\underline{\mathbf{P}}^{(\zeta)} = \underline{\mathbf{M}}^{(\zeta)} \underline{\mathbf{Q}}^{(\zeta)}$ or

$$\underline{\mathbf{M}}^{(\zeta)} = \underline{\mathbf{P}}^{(\zeta)} \underline{\mathbf{Q}}^{(\zeta) -1}. \tag{2.36}$$

This allows to calculate approximate second-order traveltimes of a specified wave at arbitrary points near a reference ray by dynamic ray tracing along that ray. If a point on the reference ray is specified by $\zeta_i$, $i = 1, 2, 3$, the second-order approximate traveltime at point $\zeta_i + \Delta \zeta_i$, $i = 1, 2, 3$, is given by

$$\tau \left( \zeta_1 + \Delta \zeta_1, \ \zeta_2 + \Delta \zeta_2, \ \zeta_3 + \Delta \zeta_3 \right) = \tau \left( \zeta_1, \zeta_2, \zeta_3 \right) + \sum_{i=1}^{3} p_i^{(\zeta)} \Delta \zeta_i + \frac{1}{2} \sum_{i,j=1}^{3} M_{ij}^{(\zeta)} \Delta \zeta_i \Delta \zeta_j, \tag{2.37}$$

where $M_{ij}^{(\zeta)} = \partial^2 \tau / \partial \zeta_i \partial \zeta_j$, $i, j = 1, 2, 3$. Evaluated at $\zeta_3$, this expression reduces to

$$\tau \left( \boldsymbol{\zeta} + \Delta \boldsymbol{\zeta}, \ \zeta_3 \right) = \tau \left( \boldsymbol{\zeta}, \zeta_3 \right) + \mathbf{p}^{(\zeta)} \cdot \Delta \boldsymbol{\zeta} + \frac{1}{2} \Delta \boldsymbol{\zeta}^T \underline{\mathbf{M}}^{(\zeta)} \Delta \boldsymbol{\zeta}, \tag{2.38}$$

where the notation $\boldsymbol{\zeta} = (\zeta_1, \zeta_2)^T$, $\Delta\boldsymbol{\zeta} = (\Delta\zeta_1, \Delta\zeta_2)^T$, and $\mathbf{p}^{(\zeta)} = (p_1^{(\zeta)}, p_2^{(\zeta)})^T$ has been used, and $\underline{\mathbf{M}}^{(\zeta)}$ is given by equation (2.36).

As noted earlier, the submatrix $(\underline{\mathbf{Q}}_2^{(\zeta)}, \underline{\mathbf{P}}_2^{(\zeta)})^T$ of the ray propagator matrix $\underline{\mathbf{\Pi}}^{(\zeta)}$ is a solution of the dynamic ray-tracing system for initial conditions $(\mathbf{0}_2, \mathbf{I}_2)^T$. These initial conditions correspond to a point source and are, therefore, also known as *normalized point-source* initial conditions (e. g., Červený, 2001). The second derivatives of the traveltime field corresponding to a point source are, thus, given by

$$\underline{\mathbf{M}}^{(\zeta)} = \underline{\mathbf{P}}_2^{(\zeta)} \, \underline{\mathbf{Q}}_2^{(\zeta)\,-1} \, . \tag{2.39}$$

## 2.4 Ray perturbation theory

Paraxial ray theory as described in Section 2.3 may be regarded as describing the first-order effect of perturbations $\Delta\boldsymbol{\eta} = (\Delta\boldsymbol{\zeta}, \Delta\mathbf{p}^{(\zeta)})^T$ of the initial parameters of a ray on its phase space trajectory $\boldsymbol{\eta}(\sigma) = (\boldsymbol{\zeta}(\sigma), \mathbf{p}^{(\zeta)}(\sigma))^T$. Paraxial ray theory is, thus, a special case of ray perturbation theory.

It is, however, also possible to describe the first-order effect of perturbations of the Hamiltonian itself (due to perturbations of the velocity $v(\zeta_1, \zeta_2, \zeta_3)$ in the medium) on the phase space trajectory $\boldsymbol{\eta}(\sigma)$ of a ray. In addition, the first-order effects of perturbing the initial parameters $\boldsymbol{\eta}(\sigma_0)$ of a ray, or the velocity $v(\zeta_1, \zeta_2, \zeta_3)$ along a ray, on the ray propagator matrix of that ray may be considered. The resulting perturbed ray propagator matrix describes paraxial rays in the vicinity of a reference ray that has been perturbed from its original trajectory due to perturbations of its initial parameters or perturbations of the Hamiltonian (caused by perturbations of the velocity along the ray).

The development of ray perturbation theory presented in this section follows the approach of Farra and Madariaga (1987). The resulting expressions play a fundamental role in the formulation of the tomographic inversion method introduced in Chapter 4.

If a smooth perturbation $\Delta v(\zeta_1, \zeta_2, \zeta_3)$ of the inhomogeneous background medium velocity $v(\zeta_1, \zeta_2, \zeta_3)$ is introduced, such that the overall velocity is described by

$$v + \Delta v \, , \tag{2.40}$$

the corresponding perturbed Hamiltonian may be written as

$$H = H_0 + \Delta H \, , \tag{2.41}$$

with

$$\Delta H = \frac{\partial H}{\partial v} \Delta v \, . \tag{2.42}$$

Inserting expressions (2.22) into the left-hand side of the Hamiltonian system (2.20), substituting a Taylor expansion for $H$ given by equation (2.41) into the right-hand side, and keeping only linear terms in the perturbations $\Delta\zeta_1$, $\Delta\zeta_2$, $\Delta p_1^{(\zeta)}$, $\Delta p_2^{(\zeta)}$, and $\Delta v$ results in

$$\frac{d\Delta\boldsymbol{\eta}}{d\sigma} = \underline{\mathbf{S}} \, \Delta\boldsymbol{\eta} + \Delta\mathbf{w} \tag{2.43}$$

19

with

$$\Delta \mathbf{w} = \left( \frac{\partial \Delta H}{\partial p_1^{(\zeta)}}, \frac{\partial \Delta H}{\partial p_2^{(\zeta)}}, -\frac{\partial \Delta H}{\partial \zeta_1}, -\frac{\partial \Delta H}{\partial \zeta_2} \right)^T . \tag{2.44}$$

This system of equations may be solved in terms of the propagator matrix $\underline{\mathbf{\Pi}}^{(\zeta)}$ to yield (e. g., Gilbert and Backus, 1966)

$$\Delta \boldsymbol{\eta}(\sigma) = \underline{\mathbf{\Pi}}^{(\zeta)}(\sigma, \sigma_0) \Delta \boldsymbol{\eta}(\sigma_0) + \int_{\sigma_0}^{\sigma} \underline{\mathbf{\Pi}}^{(\zeta)}(\sigma, \sigma') \Delta \mathbf{w}(\sigma') \, d\sigma' . \tag{2.45}$$

If only perturbations of the ray trajectory due to perturbations of the velocity are sought, the first term on the right-hand side of equation (2.45) is zero, as $\Delta \boldsymbol{\eta}(\sigma_0) = \mathbf{0}$.

For practical applications of equation (2.45), it is convenient to make use of the chain rule for propagator matrices, equation (2.32):

$$\underline{\mathbf{\Pi}}_0^{(\zeta)}(\sigma, \sigma') = \underline{\mathbf{\Pi}}_0^{(\zeta)}(\sigma, \sigma_0) \underline{\mathbf{\Pi}}_0^{(\zeta)}(\sigma_0, \sigma') = \underline{\mathbf{\Pi}}_0^{(\zeta)}(\sigma, \sigma_0) \underline{\mathbf{\Pi}}_0^{(\zeta)-1}(\sigma', \sigma_0) . \tag{2.46}$$

Substituting this expression for $\underline{\mathbf{\Pi}}_0^{(\zeta)}(\sigma, \sigma')$ in equation (2.45) allows to directly perform the integration along the ray during ray tracing, as the inverse propagator matrix may be written in terms of the elements of the forward propagator matrix, see equation (2.31).

To determine the perturbation of the propagator matrix of a ray due to perturbations of the ray's initial parameters (initial phase space coordinates), or due to velocity perturbations along the ray, the first-order effect of such perturbations on the matrix $\underline{\mathbf{S}}$, equation (2.24), needs to be considered. The perturbed matrix $\underline{\mathbf{S}}$ may be written as

$$\underline{\mathbf{S}} = \underline{\mathbf{S}}_0 + \Delta \underline{\mathbf{S}} \tag{2.47}$$

with

$$\Delta \underline{\mathbf{S}} = \Delta \underline{\mathbf{S}}_1 (\Delta v) + \Delta \underline{\mathbf{S}}_2 (\Delta \boldsymbol{\eta}) . \tag{2.48}$$

The elements of $\Delta \underline{\mathbf{S}}_1$ are obtained by substituting expression (2.41) into the definition of $\underline{\mathbf{S}}$, equation (2.24):

$$\Delta \underline{\mathbf{S}}_1 = \begin{pmatrix} \frac{\partial^2 \Delta H}{\partial p_1^{(\zeta)} \partial \zeta_1} & \frac{\partial^2 \Delta H}{\partial p_1^{(\zeta)} \partial \zeta_2} & \frac{\partial^2 \Delta H}{\partial p_1^{(\zeta)} \partial p_1^{(\zeta)}} & \frac{\partial^2 \Delta H}{\partial p_1^{(\zeta)} \partial p_2^{(\zeta)}} \\ \frac{\partial^2 \Delta H}{\partial p_2^{(\zeta)} \partial \zeta_1} & \frac{\partial^2 \Delta H}{\partial p_2^{(\zeta)} \partial \zeta_2} & \frac{\partial^2 \Delta H}{\partial p_2^{(\zeta)} \partial p_1^{(\zeta)}} & \frac{\partial^2 \Delta H}{\partial p_2^{(\zeta)} \partial p_2^{(\zeta)}} \\ -\frac{\partial^2 \Delta H}{\partial \zeta_1 \partial \zeta_1} & -\frac{\partial^2 \Delta H}{\partial \zeta_1 \partial \zeta_2} & -\frac{\partial^2 \Delta H}{\partial \zeta_1 \partial p_1^{(\zeta)}} & -\frac{\partial^2 \Delta H}{\partial \zeta_1 \partial p_2^{(\zeta)}} \\ -\frac{\partial^2 \Delta H}{\partial \zeta_2 \partial \zeta_1} & -\frac{\partial^2 \Delta H}{\partial \zeta_2 \partial \zeta_2} & -\frac{\partial^2 \Delta H}{\partial \zeta_2 \partial p_1^{(\zeta)}} & -\frac{\partial^2 \Delta H}{\partial \zeta_2 \partial p_2^{(\zeta)}} \end{pmatrix} \tag{2.49}$$

with $\Delta H = \frac{\partial H}{\partial v} \Delta v$.

The matrix $\Delta \underline{\mathbf{S}}_2$ represents the perturbation of $\underline{\mathbf{S}}$ due to perturbations of the elements of $\boldsymbol{\eta}$. It may be written as (Farra and Madariaga, 1987)

$$\Delta \underline{\mathbf{S}}_2 = \sum_{i=1}^{2} \left[ \Delta \zeta_i \frac{\partial}{\partial \zeta_i} + \Delta p_i^{(\zeta)} \frac{\partial}{\partial p_i^{(\zeta)}} \right] \begin{pmatrix} \frac{\partial^2 H_0}{\partial p_1^{(\zeta)} \partial \zeta_1} & \frac{\partial^2 H_0}{\partial p_1^{(\zeta)} \partial \zeta_2} & \frac{\partial^2 H_0}{\partial p_1^{(\zeta)} \partial p_1^{(\zeta)}} & \frac{\partial^2 H_0}{\partial p_1^{(\zeta)} \partial p_2^{(\zeta)}} \\ \frac{\partial^2 H_0}{\partial p_2^{(\zeta)} \partial \zeta_1} & \frac{\partial^2 H_0}{\partial p_2^{(\zeta)} \partial \zeta_2} & \frac{\partial^2 H_0}{\partial p_2^{(\zeta)} \partial p_1^{(\zeta)}} & \frac{\partial^2 H_0}{\partial p_2^{(\zeta)} \partial p_2^{(\zeta)}} \\ -\frac{\partial^2 H_0}{\partial \zeta_1 \partial \zeta_1} & -\frac{\partial^2 H_0}{\partial \zeta_1 \partial \zeta_2} & -\frac{\partial^2 H_0}{\partial \zeta_1 \partial p_1^{(\zeta)}} & -\frac{\partial^2 H_0}{\partial \zeta_1 \partial p_2^{(\zeta)}} \\ -\frac{\partial^2 H_0}{\partial \zeta_2 \partial \zeta_1} & -\frac{\partial^2 H_0}{\partial \zeta_2 \partial \zeta_2} & -\frac{\partial^2 H_0}{\partial \zeta_2 \partial p_1^{(\zeta)}} & -\frac{\partial^2 H_0}{\partial \zeta_2 \partial p_2^{(\zeta)}} \end{pmatrix} . \tag{2.50}$$

Again, all partial derivatives are evaluated on the central ray. The quantities $\Delta \zeta_1$, $\Delta \zeta_2$, $\Delta p_1^{(\zeta)}$, and $\Delta p_2^{(\zeta)}$ are calculated from the corresponding quantities at $\sigma_0$ by equation (2.27).

The matrix $\underline{\mathbf{S}} = \underline{\mathbf{S}}_0 + \Delta \underline{\mathbf{S}}$ allows to describe paraxial rays near the perturbed reference ray:

$$\frac{d \Delta \boldsymbol{\eta}}{d \sigma} = \left( \underline{\mathbf{S}}_0 + \Delta \underline{\mathbf{S}} \right) \Delta \boldsymbol{\eta} = \underline{\mathbf{S}}_0 \Delta \boldsymbol{\eta} + \Delta \underline{\mathbf{S}} \Delta \boldsymbol{\eta} . \tag{2.51}$$

Here, $\Delta \boldsymbol{\eta}$ denotes the perturbation of the phase space coordinates of the paraxial ray relative to the perturbed reference ray. Again, applying propagator matrix theory (e. g., Gilbert and Backus, 1966) leads to

$$\Delta \boldsymbol{\eta}(\sigma) = \underline{\boldsymbol{\Pi}}_0^{(\zeta)}(\sigma, \sigma_0) \Delta \boldsymbol{\eta}(\sigma_0) + \int_{\sigma_0}^{\sigma} \underline{\boldsymbol{\Pi}}_0^{(\zeta)}(\sigma, \sigma') \Delta \underline{\mathbf{S}}(\sigma') \Delta \boldsymbol{\eta}(\sigma') d\sigma' . \tag{2.52}$$

To first order in the perturbation $\Delta \underline{\mathbf{S}}$ (first Born approximation), equation (2.52) yields (Farra and Madariaga, 1987)

$$\Delta \boldsymbol{\eta}(\sigma) = \left[ \underline{\boldsymbol{\Pi}}_0^{(\zeta)}(\sigma, \sigma_0) + \Delta \underline{\boldsymbol{\Pi}}^{(\zeta)}(\sigma, \sigma_0) \right] \Delta \boldsymbol{\eta}(\sigma_0) \tag{2.53}$$

with

$$\Delta \underline{\boldsymbol{\Pi}}^{(\zeta)}(\sigma, \sigma_0) = \int_{\sigma_0}^{\sigma} \underline{\boldsymbol{\Pi}}_0^{(\zeta)}(\sigma, \sigma') \Delta \underline{\mathbf{S}}(\sigma') \underline{\boldsymbol{\Pi}}_0^{(\zeta)}(\sigma', \sigma_0) d\sigma' . \tag{2.54}$$

For practical applications of equation (2.54) it is again useful to substitute expression (2.46) for $\underline{\boldsymbol{\Pi}}_0^{(\zeta)}(\sigma, \sigma')$, allowing to perform the integration directly during ray tracing.

Here, the paraxial and dynamic ray-tracing systems and ray perturbation theory have been derived for general curvilinear orthogonal coordinate systems. Due to the use of a reduced Hamiltonian, equation (2.18), one of the coordinates acts as the running parameter $\sigma$ and must therefore vary monotonously along the ray. Paraxial and dynamic ray tracing as well as ray perturbation theory may also be formulated without this restriction by avoiding the use of a reduced Hamiltonian and directly using $\tilde{H} \left( \zeta_1, \zeta_2, \zeta_3, p_1^{(\zeta)}, p_2^{(\zeta)}, p_3^{(\zeta)} \right) = 0$ (e. g., Červený, 2001). However, this increases the number of equations to be solved.

In the following sections, the theory formulated in general curvilinear orthogonal coordinates will be applied to two different coordinate systems, which will then be used in the course of later chapters in the context of tomographic inversion. The first one is the ray-centered coordinate system

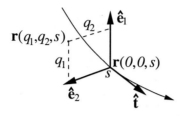

Figure 2.1: Definition of ray-centered coordinates. See text for details.

(e. g., Popov and Pšenčík, 1978; Červený and Hron, 1980; Popov, 2002) in which many of the expressions given in the previous sections simplify considerably. The second one is the Cartesian coordinate system with the $x_3$-coordinate used as the running parameter along the ray. While the use of the $x_3$-coordinate as the running parameter allows only wave propagation along central rays that have no turning point with respect to the $x_3$ direction to be described, this formulation leads to a number of numerical advantages.

## 2.5   Ray-centered coordinates

In this section, the ray-centered coordinate system (e. g., Popov and Pšenčík, 1978; Popov, 2002; Červený, 2001) will be introduced and the results of the previous sections will be specialized to this coordinate system.

As a starting point, consider a given ray in a smooth medium with a known trajectory, determined, for example, with the ray-tracing system (2.11). Let $s$ be the arclength along the ray from some reference point. At any point $s$ along that ray, called the central ray, two mutually orthogonal unit vectors $\hat{\mathbf{e}}_1(s)$ and $\hat{\mathbf{e}}_2(s)$ can be introduced in the plane normal to the ray through $s$. The behavior of these two unit vectors along the ray is described by the following differential equations:

$$\frac{d\hat{\mathbf{e}}_i}{ds} = \kappa_i(s)\hat{\mathbf{t}}(s) \qquad i = 1, 2 , \tag{2.55}$$

where $\kappa_1(s)$ and $\kappa_2(s)$ have yet to be specified, and $\hat{\mathbf{t}}(s)$ is a unit vector tangent to the ray at $s$. A point in the vicinity of the ray may then be described by the three coordinates $(q_1, q_2, s)$ as

$$\mathbf{r}(q_1, q_2, s) = q_1\hat{\mathbf{e}}_1(s) + q_2\hat{\mathbf{e}}_2(s) + \mathbf{r}(0, 0, s) , \tag{2.56}$$

see Figure 2.1. The infinitesimal line element $dS$ in ray-centered coordinates is given by

$$dS^2 = d\mathbf{r} \cdot d\mathbf{r} = dq_1^2 + dq_2^2 + h^2 ds^2 , \tag{2.57}$$

where $\hat{\mathbf{t}}(s) = d\mathbf{r}(0, 0, s)/ds$ has been used and

$$h = 1 + \kappa_1(s)q_1 + \kappa_2(s)q_2 . \tag{2.58}$$

Thus, the introduced coordinate system is orthogonal and regular in some vicinity around the central ray. The region of regularity depends on the curvature of the ray. From (2.58) it also follows that the eikonal equation in ray-centered coordinates is given by, compare equation (2.15),

$$p_1^2 + p_2^2 + h^{-2} p_s^2 = v^{-2} \tag{2.59}$$

with $p_1 = \partial\tau/\partial q_1$, $p_2 = \partial\tau/\partial q_2$, and $p_s = \partial\tau/\partial s$. The reduced Hamiltonian (2.18) then reads

$$H = -h\sqrt{v^{-2} - p_1^2 - p_2^2} = -p_s \ . \tag{2.60}$$

Following the procedures in Section 2.2, the ray-tracing system can be set up. However, as the ray-centered coordinate system is attached to an already known ray which in this coordinate system has coordinates $q_1 = q_2 \equiv 0$ and $p_1 = p_2 \equiv 0$, the right-hand side of equation (2.16) must be identically zero. From this condition, $\kappa_1(s)$ and $\kappa_2(s)$ can be obtained:

$$\kappa_i(s) = \frac{1}{V} V_i \qquad i = 1, 2 \ , \tag{2.61}$$

where $V = v\big|_{(0,0,s)}$ and $V_i = \frac{\partial v}{\partial q_i}\big|_{(0,0,s)}$. The expression for the scaling factor $h$ then becomes

$$h = \left[1 + V^{-1}(V_1 q_1 + V_2 q_2)\right] \ . \tag{2.62}$$

The paraxial ray-tracing system can be obtained from $H$ given in equation (2.60) as described in Section 2.3. If the notation $\Delta\boldsymbol{\eta} = \left(\Delta q_1, \Delta q_2, \Delta p_1, \Delta p_2\right)^T$ is used, the paraxial ray-tracing system reads

$$\frac{d\Delta\boldsymbol{\eta}}{ds} = \underline{\mathbf{S}}\Delta\boldsymbol{\eta} \tag{2.63}$$

with

$$\underline{\mathbf{S}} = \begin{pmatrix} \mathbf{0}_2 & V\,\mathbf{I}_2 \\ -V^{-2}\underline{\mathbf{V}} & \mathbf{0}_2 \end{pmatrix} \ . \tag{2.64}$$

Here, $\underline{\mathbf{V}}$ is a $2 \times 2$ matrix with $V_{ij} = \frac{\partial^2 v}{\partial q_i \partial q_j}\big|_{(0,0,s)}$. The associated ray propagator matrix is denoted by

$$\underline{\mathbf{\Pi}}(s, s_0) = \begin{pmatrix} \underline{\mathbf{Q}}_1(s, s_0) & \underline{\mathbf{Q}}_2(s, s_0) \\ \underline{\mathbf{P}}_1(s, s_0) & \underline{\mathbf{P}}_2(s, s_0) \end{pmatrix} \ , \tag{2.65}$$

where again $\left(\underline{\mathbf{Q}}_1, \underline{\mathbf{P}}_1\right)^T$ is a solution of the dynamic ray-tracing system for initial conditions $\left(\mathbf{I}_2, \mathbf{0}_2\right)^T$, which are in the case of ray-centered coordinates known as *normalized plane-wave* initial conditions (e. g., Červený, 2001), as rays with $\partial p_i/\partial\gamma_j = 0$ at $s_0$ are initially parallel to the central ray, and are thus associated with a plane wave. As noted earlier, $\left(\mathbf{0}_2, \mathbf{I}_2\right)^T$ at $s_0$ corresponds to normalized point-source initial conditions.

Just like in the general case described in earlier sections, the second derivatives of traveltime with respect to the coordinates $q_1$ and $q_2$ at constant $s$ for a specified wave can be obtained from the solution of the dynamic ray-tracing system (2.34):

$$\underline{\mathbf{M}} = \underline{\mathbf{P}}\underline{\mathbf{Q}}^{-1} \ . \tag{2.66}$$

Due to the fact that in ray-centered coordinates, $p_1 = p_2 = 0$ on the central ray, the expression for the second-order approximate paraxial traveltimes of a system of rays near the central ray, equation (2.38), takes a particularly simple form. Using the notation $\Delta\mathbf{q} = (\Delta q_1, \Delta q_2)^T$, it reads

$$\tau(\Delta\mathbf{q}, s) = \tau(0, 0, s) + \frac{1}{2}\Delta\mathbf{q}^T\underline{\mathbf{M}}\Delta\mathbf{q} \ . \tag{2.67}$$

It is shown, among others, by Červený (2001), that the matrix $\underline{\mathbf{M}}$ of second derivatives of traveltime in ray-centered coordinates is related in a simple way to the matrix of wavefront curvature $\underline{\mathbf{K}}$ at the location $s$ on the central ray:

$$\underline{\mathbf{M}} = \frac{1}{v}\underline{\mathbf{K}} \ , \tag{2.68}$$

or, introducing the matrix $\underline{\mathbf{R}} = \underline{\mathbf{K}}^{-1}$ of radii of wavefront curvature:

$$\underline{\mathbf{M}} = (v\underline{\mathbf{R}})^{-1} \ . \tag{2.69}$$

The matrices $\underline{\mathbf{M}}$, $\underline{\mathbf{K}}$, or $\underline{\mathbf{R}}$, respectively, may also be determined directly by solving a nonlinear first-order ordinary differential equation of Riccati type derived from the dynamic ray-tracing system (e. g., Červený, 2001).

The expressions of ray perturbation theory in general curvilinear orthogonal coordinates given in Section 2.4 may be specialized to the case of ray-centered coordinates. However, for the tomographic inversion described in the following chapters, ray-centered coordinates will only be applied in the two-dimensional case. The corresponding ray perturbation theory results in 2D are given in Appendix D.

### 2.5.1 Ray-centered coordinates in the 2D case

In the case of wave propagation confined to a 2D plane, the expressions for ray tracing in ray-centered coordinates simplify considerably. This may occur, for instance, in media which are invariant in one spatial direction, say, the $x_2$-direction. The trajectories of rays propagating in the $x_1$-$x_3$-plane may then be described with two spatial coordinates only.

Consequently, only two ray-centered coordinates $(q, s)$ remain, where $s$ is still the arclength along the ray and $q$ corresponds to the direction normal to the ray. The reduced Hamiltonian for the case of 2D ray-centered coordinates is given by

$$H = -h\sqrt{v^{-2} - p^2} \ , \tag{2.70}$$

where $p = \frac{\partial\tau}{\partial q}$ and

$$h = \left[1 + \left(\frac{1}{v}\frac{\partial v}{\partial q}\right)\Big|_{q=0} q\right] \ . \tag{2.71}$$

The paraxial ray-tracing system then reads

$$\frac{d}{ds}\begin{pmatrix} \Delta q \\ \Delta p \end{pmatrix} = \underline{\mathbf{S}}\begin{pmatrix} \Delta q \\ \Delta p \end{pmatrix} \quad \text{with} \quad \underline{\mathbf{S}} = \begin{pmatrix} 0 & v \\ -v^{-2}\frac{\partial^2 v}{\partial q^2} & 0 \end{pmatrix} \ , \tag{2.72}$$

where all elements of $\underline{S}$ are evaluated on the central ray. The associated $2 \times 2$ ray propagator matrix is given by

$$\underline{\Pi}(s,s_0) = \begin{pmatrix} Q_1(s,s_0) & Q_2(s,s_0) \\ P_1(s,s_0) & P_2(s,s_0) \end{pmatrix}, \tag{2.73}$$

where $Q_1$, $Q_2$, $P_1$, and $P_2$ are scalars. The second derivative of traveltime with respect to $q$ for a specified wave, evaluated on the central ray, is

$$\frac{\partial^2 \tau}{\partial q^2} = M = \frac{P}{Q}. \tag{2.74}$$

In particular, for the case of a point source at $s = s_0$,

$$M = \frac{P_2}{Q_2}. \tag{2.75}$$

The associated (scalar) wavefront curvature $K$ and radius of wavefront curvature $R$ are given by

$$M = \frac{1}{v}K = \frac{1}{vR}. \tag{2.76}$$

As noted above, the ray perturbation theory expressions for 2D ray-centered coordinates can be found in Appendix D.

## 2.6 Cartesian coordinates

In Cartesian coordinates $(x_1, x_2, x_3)$, the scaling factors $h_1$, $h_2$, and $h_3$ are equal to one. The eikonal equation, thus, simply reads

$$(p_1^{(x)})^2 + (p_2^{(x)})^2 + (p_3^{(x)})^2 = \frac{1}{v^2(x_1, x_2, x_3)} \tag{2.77}$$

with $p_i^{(x)} = \partial \tau / \partial x_i$, $i = 1, 2, 3$. If wave propagation occurs predominantly in the $x_3$-direction, the coordinate $x_3$ may be used as the free parameter along the ray. The reduced Hamiltonian (2.18) is then given by

$$H = -\sqrt{v^{-2} - (p_1^{(x)})^2 - (p_2^{(x)})^2} = -p_3^{(x)}. \tag{2.78}$$

The associated reduced ray-tracing system reads

$$\frac{dx_1}{dx_3} = \frac{p_1^{(x)}}{p_3^{(x)}}, \qquad \frac{dp_1^{(x)}}{dx_3} = -\frac{1}{v^3 p_3^{(x)}} \frac{\partial v}{\partial x_1},$$

$$\frac{dx_2}{dx_3} = \frac{p_2^{(x)}}{p_3^{(x)}}, \qquad \frac{dp_2^{(x)}}{dx_3} = -\frac{1}{v^3 p_3^{(x)}} \frac{\partial v}{\partial x_2}, \tag{2.79}$$

and the traveltime along the ray can be obtained by integrating

$$\frac{\partial \tau}{\partial x_3} = \frac{1}{v^2 p_3^{(x)}} . \tag{2.80}$$

If, again, a vector $\Delta \boldsymbol{\eta} = \left( \Delta x_1, \Delta x_2, \Delta p_1^{(x)}, \Delta p_2^{(x)} \right)^T$ is defined, the paraxial ray-tracing system can be written as

$$\frac{d\Delta \boldsymbol{\eta}}{dx_3} = \underline{S} \Delta \boldsymbol{\eta} , \tag{2.81}$$

where the elements of the $4 \times 4$ matrix $\underline{S}$ can be calculated from (2.24) with the reduced Hamiltonian $H$ defined in (2.78):

$$S_{11} = \frac{p_1^{(x)}}{v^3 \left( p_3^{(x)} \right)^3} \frac{\partial v}{\partial x_1} , \qquad\qquad S_{13} = \frac{1}{p_3^{(x)}} + \frac{\left( p_1^{(x)} \right)^2}{\left( p_3^{(x)} \right)^3} ,$$

$$S_{12} = \frac{p_1^{(x)}}{v^3 \left( p_3^{(x)} \right)^3} \frac{\partial v}{\partial x_2} , \qquad\qquad S_{14} = \frac{p_1^{(x)} p_2^{(x)}}{\left( p_3^{(x)} \right)^3} ,$$

$$S_{21} = \frac{p_2^{(x)}}{v^3 \left( p_3^{(x)} \right)^3} \frac{\partial v}{\partial x_1} , \qquad\qquad S_{23} = \frac{p_1^{(x)} p_2^{(x)}}{\left( p_3^{(x)} \right)^3} ,$$

$$S_{22} = \frac{p_2^{(x)}}{v^3 \left( p_3^{(x)} \right)^3} \frac{\partial v}{\partial x_2} , \qquad\qquad S_{24} = \frac{1}{p_3^{(x)}} + \frac{\left( p_2^{(x)} \right)^2}{\left( p_3^{(x)} \right)^3} ,$$

$$S_{31} = -\frac{1}{v^4 p_3^{(x)}} \left[ v \frac{\partial^2 v}{\partial x_1^2} + \left( \frac{1}{v^2 \left( p_3^{(x)} \right)^2} \right) \left( \frac{\partial v}{\partial x_1} \right)^2 \right] , \qquad S_{33} = -\frac{p_1^{(x)}}{v^3 \left( p_3^{(x)} \right)^3} \frac{\partial v}{\partial x_1} ,$$

$$S_{32} = -\frac{1}{v^4 p_3^{(x)}} \left[ v \frac{\partial^2 v}{\partial x_1 \partial x_2} + \left( \frac{1}{v^2 \left( p_3^{(x)} \right)^2} \right) \left( \frac{\partial v}{\partial x_1} \right) \left( \frac{\partial v}{\partial x_2} \right) \right] , \qquad S_{34} = -\frac{p_2^{(x)}}{v^3 \left( p_3^{(x)} \right)^3} \frac{\partial v}{\partial x_1} ,$$

$$S_{41} = -\frac{1}{v^4 p_3^{(x)}} \left[ v \frac{\partial^2 v}{\partial x_1 \partial x_2} + \left( \frac{1}{v^2 \left( p_3^{(x)} \right)^2} \right) \left( \frac{\partial v}{\partial x_1} \right) \left( \frac{\partial v}{\partial x_2} \right) \right] , \qquad S_{43} = -\frac{p_1^{(x)}}{v^3 \left( p_3^{(x)} \right)^3} \frac{\partial v}{\partial x_2} ,$$

$$S_{42} = -\frac{1}{v^4 p_3^{(x)}} \left[ v \frac{\partial^2 v}{\partial x_2^2} + \left( \frac{1}{v^2 \left( p_3^{(x)} \right)^2} \right) \left( \frac{\partial v}{\partial x_2} \right)^2 \right] , \qquad S_{44} = -\frac{p_2^{(x)}}{v^3 \left( p_3^{(x)} \right)^3} \frac{\partial v}{\partial x_2} .$$

The associated $4 \times 4$ ray propagator matrix will be denoted by

$$\underline{\Pi}^{(x)} = \begin{pmatrix} \underline{Q}_1^{(x)} & \underline{Q}_2^{(x)} \\ \underline{P}_1^{(x)} & \underline{P}_2^{(x)} \end{pmatrix} . \tag{2.82}$$

As in the general case, solutions $\left( \mathbf{Q}^{(x)}, \underline{P}^{(x)} \right)^T$ of the dynamic ray-tracing system may be used to calculate the second derivatives of traveltime of a system of rays with respect to the coordinates $x_1$ and $x_2$:

$$\underline{M}^{(x)} = \underline{P}^{(x)} \underline{Q}^{(x)\,-1} \tag{2.83}$$

with $M_{11} = \partial^2\tau/\partial x_1^2$, $M_{12} = M_{21} = \partial^2\tau/\partial x_1\partial x_2$, and $M_{22} = \partial^2\tau/\partial x_2^2$. Paraxial traveltimes of rays near the central ray belonging to a specified wave are then given by

$$\tau(x_1 + \Delta x_1, x_2 + \Delta x_2, x_3) = \tau(x_1, x_2, x_3) + \sum_{i=1}^{2} p_i^{(x)}\Delta x_i + \sum_{i,j=1}^{2} M_{ij}^{(x)}\Delta x_i \Delta x_j \,. \tag{2.84}$$

In particular, the second derivatives of traveltime for a point source at the initial point on the central ray are given by

$$\underline{\mathbf{M}}^{(x)} = \underline{\mathbf{P}}_2^{(x)}\underline{\mathbf{Q}}_2^{(x)\,-1} \,. \tag{2.85}$$

The ray perturbation theory expressions for the case of Cartesian coordinates discussed here, are given in Appendix E.

## 2.7 Transformation from ray-centered to Cartesian coordinates

In this section it will be shown, how the second-order approximate traveltime field associated with a specified wave can be transformed from ray-centered coordinates to global Cartesian coordinates. Such a transformation will be required in later chapters in the context of the common-reflection-surface stack and the tomographic inversion with kinematic wavefield attributes. In ray-centered coordinates, the traveltime at a point $P$ with coordinates $(\Delta q_1, \Delta q_2, s')$ may up to second order be expressed in terms of quantities at a point $P_0$ on the central ray with coordinates $(0, 0, s)$ by (Červený, 2001)

$$\tau(\Delta q_1, \Delta q_2, s') = \tau(0, 0, s) + v^{-1}(s)\Delta s - \frac{1}{2}v^{-2}(s)\frac{\partial v}{\partial s}\Delta s^2 + \frac{1}{2}\sum_{i,j=1}^{2} M_{ij}\Delta q_i \Delta q_j \,, \tag{2.86}$$

where $\Delta s = (s' - s)$ and $\underline{\mathbf{M}}$ is the $2 \times 2$ matrix of second traveltime derivatives with respect to $q_1$ and $q_2$ defined in Section 2.5. The elements of $\underline{\mathbf{M}}$ in equation (2.86) are evaluated at point $P_0$.

The same traveltime field may be expressed in terms of a local ray-centered Cartesian coordinate system $(y_1, y_2, y_3)$ with its origin on the central ray at $s$ (Figure 2.2a). If the $y_1$ and $y_2$ axes coincide with the $q_1$ and $q_2$ axes, while the $y_3$ axis is tangential to the ray at $s$, the traveltime field at point $P$ is given by

$$\tau(y_1, y_2, y_3) = \tau(0, 0, 0) + v^{-1}y_3 + \frac{1}{2}\sum_{i,j=1}^{3} M_{ij}^{(y)}y_i y_j \,, \tag{2.87}$$

where the $3 \times 3$ matrix $\hat{\underline{\mathbf{M}}}^{(y)}$ with elements $M_{ij}$ has the form

$$\hat{\underline{\mathbf{M}}}^{(y)} = \begin{pmatrix} M_{11} & M_{12} & -\frac{1}{v^2}\frac{\partial v}{\partial y_1} \\ M_{21} & M_{22} & -\frac{1}{v^2}\frac{\partial v}{\partial y_2} \\ -\frac{1}{v^2}\frac{\partial v}{\partial y_1} & -\frac{1}{v^2}\frac{\partial v}{\partial y_2} & -\frac{1}{v^2}\frac{\partial v}{\partial y_3} \end{pmatrix} \,. \tag{2.88}$$

The upper left $2 \times 2$ submatrix of $\hat{\underline{\mathbf{M}}}^{(y)}$ with the elements $M_{11}, M_{12}, M_{21}$, and $M_{22}$ is identical to the

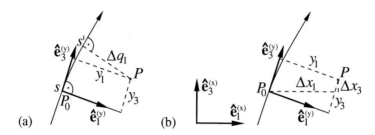

Figure 2.2: (a) Transformation from ray-centered coordinates to local ray-centered Cartesian co-ordinates at point $P_0$. (b) Transformation from local ray-centered Cartesian to global Cartesian coordinates. For simplicity, only a transformation in two dimensions is displayed. See text for details.

matrix $\underline{\mathbf{M}}$ defined in Section 2.5. The derivatives of velocity are taken at point $P_0 = (0,0,0)$. The linear term in (2.87) may be interpreted as $\sum_{i=1}^{3} p_i^{(y)} y_i$ with $p_1^{(y)} = p_2^{(y)} = 0$, and $p_3^{(y)} = v^{-1}$. In order to transform this expression from local ray-centered Cartesian coordinates to global Cartesian coordinates $(x_1, x_2, x_3)$, Figure 2.2b, a $3 \times 3$ orthonormal transformation matrix $\hat{\mathbf{H}}$ with $H_{ij} = \frac{\partial x_i}{\partial y_j}$ is required. If the origin of the local Cartesian coordinate system (point $P_0$) has the global Cartesian coordinates $(x_{01}, x_{02}, x_{03})^T$, the coordinates $(x_1, x_2, x_3)^T$ of the point $P$ are determined by

$$\Delta x_i = (x_i - x_{0i}) = \sum_{j=1}^{3} H_{ij} y_j \qquad i = 1, 2, 3 \ . \tag{2.89}$$

Correspondingly, the slowness vector $(p_1^{(x)}, p_2^{(x)}, p_3^{(x)})$ at $(x_{01}, x_{02}, x_{03})^T$ can be calculated from

$$p_i^{(x)} = \sum_{i,j=1}^{3} H_{ij} p_j^{(y)} \tag{2.90}$$

and the $3 \times 3$ matrix $\hat{\mathbf{M}}^{(x)}$ of second derivatives of traveltime with elements $M_{ij}^{(x)} = \frac{\partial^2 \tau}{\partial x_i \partial x_j}$ is obtained from

$$\hat{\mathbf{M}}^{(x)} = \hat{\mathbf{H}} \hat{\mathbf{M}}^{(y)} \hat{\mathbf{H}}^T \ . \tag{2.91}$$

The resulting paraxial traveltime expression in global Cartesian coordinates reads

$$\tau(x_1, x_2, x_3) = \tau(x_{01}, x_{02}, x_{03}) + \sum_{i=1}^{3} p_i^{(x)} \Delta x_i + \frac{1}{2} \sum_{i,j=1}^{3} M_{ij}^{(x)} \Delta x_i \Delta x_j \ . \tag{2.92}$$

This expression simplifies if it is evaluated at $x_3 = x_{03}$ and if, additionally, the velocity is locally constant. Then, the third row and column of matrix $\hat{\mathbf{M}}^{(y)}$ in equation (2.88) are zero and the traveltime expression becomes (compare equation (2.84))

$$\tau(x_1, x_2, x_{03}) = \tau(x_{01}, x_{02}, x_{03}) + \sum_{i=1}^{2} p_i^{(x)} \Delta x_i + \frac{1}{2} \sum_{i,j=1}^{2} M_{ij}^{(x)} \Delta x_i \Delta x_j \tag{2.93}$$

with

$$\underline{\mathbf{M}}^{(x)} = \underline{\mathbf{H}}\,\underline{\mathbf{M}}\,\underline{\mathbf{H}}^{T} .$$

(2.94)

The matrix $\underline{\mathbf{H}}$ is the upper left $2 \times 2$ submatrix of $\hat{\underline{\mathbf{H}}}$, and $\underline{\mathbf{M}}$ is the matrix of second traveltime derivatives in ray-centered coordinates introduced in Section 2.5. Equation (2.93) will be useful in later chapters for the description of the second-order traveltimes of emerging wavefronts at a plane measurement surface.

# Chapter 3

# The common-reflection-surface stack

In this chapter, the common-reflection-surface (CRS) stack method (e. g., Mann et al., 1999; Jäger et al., 2001; Mann, 2002) will be introduced. The CRS stack makes use of the redundancy of seismic multicoverage data to obtain a stacked simulated zero-offset section (2D case) or volume (3D case) with an improved signal-to-noise (S/N) ratio. At the same time, the process extracts traveltime information from the data in the form of a number of so-called kinematic wavefield attributes assigned to each considered zero-offset sample. These kinematic wavefield attributes form the basis of the tomographic inversion method that will be presented in Chapter 4. The CRS stack is based on a second-order traveltime approximation and can be seen as a generalization of the well-known common-midpoint (CMP) stack technique.

## 3.1 Seismic multicoverage data

As discussed in Section 1.1, reflection seismic data are usually recorded in a way that allows reflectors in the subsurface to be illuminated by multiple experiments with varying source-receiver separation (offset), thereby providing redundant information on subsurface structures. The redundancy in such multicoverage data can be used for a number of purposes. The signal-to-noise (S/N) ratio may be improved by summing (stacking) signals associated with the same reflection point in the subsurface recorded with varying source-receiver separation (offset). In addition, the variation of traveltimes of reflection events associated with a reflection point in the subsurface as a function of source-receiver offset contains information on the distribution of seismic velocities in the subsurface. In fact, this offset-dependence of reflection traveltimes is the only information available (apart form borehole measurements and geological a priori knowledge) for the construction of a velocity model, required for transforming the measured data into a structural image in the depth domain. Different methods of using that information for velocity model building are discussed in Section 1.3. The problem, however, is that it is initially unknown where exactly the signals associated with a common reflection point in the subsurface can be found in the data. In order to tackle this problem, a number of approximations and simplifying assumption about subsurface structures are often made. A well-known and frequently used method based on such simplifying assumptions is the common-midpoint (CMP) stack technique (Mayne, 1962) discussed in Section 3.2.

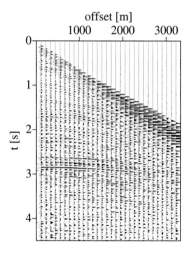

Figure 3.1: A common-midpoint (CMP) gather from a marine seismic dataset. Reflection events are visible across several offsets.

For the discussion of the CMP stack and later also of the CRS stack and the corresponding travel-time approximations, it is convenient to introduce new coordinates, called midpoint and half-offset coordinates. Throughout this chapter, a plane measurement surface is assumed with points on this surface defined by two-component vectors $\boldsymbol{\xi}$. If a source location on the measurement surface is given by $\boldsymbol{\xi}_s$ and a receiver location is given by $\boldsymbol{\xi}_g$, the half-offset $\mathbf{h}$ and midpoint $\boldsymbol{\xi}_m$ vectors are defined by

$$\mathbf{h} = (\boldsymbol{\xi}_g - \boldsymbol{\xi}_s)/2 \quad \text{and} \quad \boldsymbol{\xi}_m = (\boldsymbol{\xi}_g + \boldsymbol{\xi}_s)/2. \tag{3.1}$$

If the seismic data acquisition is restricted to a single straight line (2D case), these vectors reduce to scalars $h$ and $\xi_m$.

## 3.2 The common-midpoint (CMP) stack

The CMP stack makes use of the redundancy in seismic multicoverage data by considering seismic traces associated with a common midpoint but varying offsets. Signals in such CMP gathers (see Figure 3.1 for an example) are summed (stacked) coherently in the offset direction along appropriate stacking curves, resulting in stacked traces with an improved S/N ratio which are assigned to the respective midpoint locations. Under certain conditions, the stacked section thus obtained may then be interpreted as a simulated zero-offset section. The method was originally introduced by Mayne (1962) under the assumption of a horizontally layered medium, in which case reflection events measured on different traces in a CMP gather stem from a common reflection point in the subsurface located directly under the CMP location.

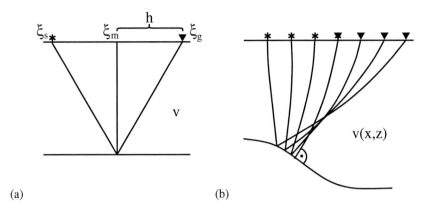

Figure 3.2: (a) A plane reflector in a homogeneous medium. In this case, reflection traveltimes in a CMP gather are exactly described by equation (3.2). (b) A curved, dipping reflector in a smooth, laterally inhomogeneous medium. To second order in the offset coordinate, reflection traveltimes in a CMP gather can be described by equation (3.4).

Consider the case of a horizontal reflector in a homogeneous medium, Figure 3.2a. The reflection traveltimes $t(h)$ measured in a CMP gather can be described by the simple expression

$$t^2(h) = t_0^2 + \frac{4h^2}{v^2} \ , \tag{3.2}$$

where $v$ is the constant medium velocity and $t_0$ is the traveltime measured at the coincident source-receiver pair at the CMP (the zero-offset traveltime). If the subsurface is no longer assumed to be homogeneous, but consists of a stack of horizontal layers with constant velocity in each layer, an equation of the form of (3.2) may still be used to describe reflection traveltimes in a CMP gather. However, it is then only a second-order approximation of the exact traveltime curve $t^2(h)$ (e. g., Taner and Koehler, 1969). Also, $v$ in (3.2) must be replaced by the root-mean-square velocity $v_{\mathrm{RMS}}$ defined by

$$v_{\mathrm{RMS}}^2 = \frac{1}{t_0} \sum_{i=1}^{N} v_i^2 \Delta t_i \ , \tag{3.3}$$

where $v_i$ is the interval velocity and $\Delta t_i$ is the two-way vertical traveltime in the $i$th layer.

In the more general case of curved, dipping reflectors in a smooth, laterally inhomogeneous medium (Figure 3.2b), or in a model consisting of layers with smooth velocity variations, separated by curved interfaces, an equation of the form (3.2) may still be used to describe reflection traveltimes in a CMP gather to second order in $h$:

$$t^2(h) = t_0^2 + \frac{4h^2}{v_{\mathrm{NMO}}^2} \ , \tag{3.4}$$

where the parameter $v_{\mathrm{NMO}}$ is called normal-moveout velocity. The absence of a linear term in equation (3.4) is due to the fact that traveltimes are invariant with respect to interchanging source

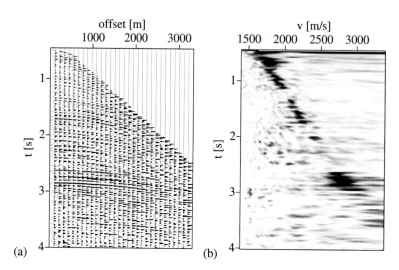

Figure 3.3: Stacking velocity analysis in a CMP gather. (a) Muted CMP gather (b) velocity spectrum. Semblance values, equation (3.5), obtained along curves defined by equation (3.4) are plotted as a function of zero-offset traveltime and stacking velocity. Maxima (dark) correspond to reflection events in (a).

and receiver locations (principle of reciprocity), thus $t(h) = t(-h)$. Note that reflections measured in a CMP gather in the general case no longer strictly correspond to a common-reflection point (CRP) in the subsurface (*reflection point dispersal*). Also, the relation of $v_{NMO}$ to the properties of the subsurface can in general not be written in a simple form. Hubral and Krey (1980) give closed-form expressions for $v_{NMO}$ as a function of azimuth for the case of a 3D inhomogeneous medium consisting of constant-velocity layers separated by curved interfaces. In smoothly varying laterally inhomogeneous media, the ray theory results of Chapter 2 together with equation (3.13), below, can be used to relate $v_{NMO}$ to the velocity distribution in the model. A closed-form expression can, however, in general not be found.

In order to use equation (3.4) for CMP stacking, appropriate values for $v_{NMO}$ need to be determined for all reflection events in a CMP gather. The process of determining velocities for stacking based on equation (3.4) is called stacking velocity analysis. A commonly used tool for performing stacking velocity analysis is the so-called velocity spectrum (e. g., Taner and Koehler, 1969; Yilmaz, 2001). To obtain the velocity spectrum of a CMP gather, a range of velocity values is used to define a set of traveltime curves for each zero-offset time $t_0$ via equation (3.4), along which a coherence analysis is carried out. For each value of $t_0$ and each tested velocity value, the obtained coherence, based on some coherence measure, is plotted (Figure 3.3). Traveltime curves which fit well to actual reflection events in the data yield high coherence values. The corresponding velocities are usually picked at a number of locations in the velocity spectrum and are interpolated to obtain a stacking velocity model. A commonly used measure of coherence is semblance given by

$$C_S = \frac{1}{N} \frac{\sum_t \left( \sum_{i=1}^N f_{i,t(i)} \right)^2}{\sum_t \sum_{i=1}^N f_{i,t(i)}^2} .$$  (3.5)

Here, $f_{i,t(i)}$ is the amplitude on the $i$th trace at two-way traveltime $t(i)$ and $N$ is the number of traces considered in the calculation of semblance. The summation over time is carried out in a small window centered about the traveltime defined by the considered traveltime curve. A number of other coherence measures can be found in Taner and Koehler (1969).

Once the stacking velocity as a function of $t_0$ has been determined, it can be used as the normal-moveout velocity $v_{NMO}$ in equation (3.4) to apply a normal-moveout correction

$$\Delta t_{NMO} = t_0 \left( \sqrt{1 + \frac{4h^2}{v_{NMO}^2 t_0^2}} - 1 \right)$$  (3.6)

to all samples in a CMP gather. Amplitudes in the resulting NMO-corrected CMP gather can then be stacked along the offset axis to obtain a stacked trace.

In practice, stacking velocities determined by velocity analysis do not coincide with normal-moveout velocities $v_{NMO}$ defined by the second-order traveltime approximation (3.4). The departure of stacking velocities from normal-moveout velocities, or more generally the departure of data-derived moveout parameters from the corresponding coefficients in a second-order traveltime approximation, is called *spread-length bias* (e. g., Al-Chalabi, 1973; Hubral and Krey, 1980). It is caused by a number of different factors, some of which are discussed in Section 6.1. The most important ones are the departure of the actual reflection traveltime curve from a hyperbolic shape due to lateral inhomogeneities in the subsurface and the finite offset aperture used during velocity analysis. Spread-length bias plays a role whenever the moveout parameters determined from the seismic data are used for further applications and are related to subsurface properties directly on the zero-offset ray, as is done, for instance, in the inversion methods discussed in Section 4.2. For such applications, the maximum considered offset in the CMP gather should be chosen with care.

In order to explicitly account for the effects of reflector dip, that is, remove reflection point dispersal and handle conflicting dips, an additional correction called dip moveout (DMO) correction is usually applied to the moveout corrected traces before stacking. After DMO correction, events in a CMP gather correspond to common-reflection points in the subsurface and it is justified to regard the resulting stacked section as a simulated zero-offset section. Here, the DMO process will not be further discussed. Details on DMO can for example be found in Deregowski (1986).

## 3.3  Basic concepts of the CRS stack

The CMP stack described in the previous section makes use of a traveltime approximation that is of second order in the half-offset coordinate. The concept of using second-order traveltime approximations for stacking can be generalized to include also the midpoint coordinate $\xi_m$. The stacking operator for 2D seismic data is then no longer a trajectory in time–midpoint–half-offset space, as defined by equation (3.4), but an entire stacking surface, which extends not only into the offset, but also into the midpoint direction. This implies the assumption that the time-domain reflection/diffraction response of subsurface structures is locally continuous across several traces in the midpoint direction, which is in accordance with observations and, in fact, forms the basis of the entire reflection seismic method.

If, like in the CMP stack, a second-order approximation of the squared traveltime around a zero-offset point $(t_0, \xi_0)$ is used, one obtains a stacking operator of the form (e. g., Schleicher et al., 1993)

$$t^2(\xi_m, h) = \left(t_0 + 2\,p^{(\xi)}\Delta\xi\right)^2 + 2t_0\left(M_N^{(\xi)}\Delta\xi^2 + M_{NIP}^{(\xi)}h^2\right) \tag{3.7}$$

with $\Delta\xi = \xi_m - \xi_0$. The meaning of the notations $M_N^{(\xi)}$ and $M_{NIP}^{(\xi)}$ for the second-order coefficients in equation (3.7) will become obvious in Section 3.4, below. Utilizing an approximation of $t^2$, rather than of $t$ itself can be justified by the fact that the approximation of $t^2$ is exact for the case of a planar reflector in a homogeneous medium. The second-order approximation of $t$, on the other hand, is in no case exact. In addition, numerical investigations by Ursin (1982) indicate that second-order traveltime approximations of $t^2$ tend to be more accurate than second-order approximations of $t$ in the case of inhomogeneous media.

The basic idea of the CRS stack method (e. g., Mann et al., 1999; Jäger et al., 2001) is to use a traveltime approximation of the form of (3.7) as a stacking operator to coherently stack reflection amplitudes in the multicoverage data in the vicinity of each zero-offset sample $(t_0, \xi_0)$, thus obtaining a stacked simulated zero-offset section (see Figure 3.4). The shape of the traveltime surface defined by equation (3.7) is controlled by the three parameters $p^{(\xi)}$, $M_N^{(\xi)}$, and $M_{NIP}^{(\xi)}$. During the CRS stack process, optimum values for these parameters are automatically determined independently for each zero-offset sample $(t_0, \xi_0)$ to be simulated. This is realized by varying the parameter values (and, thus, the operator shape) and performing a coherence analysis along the stacking operator in the multicoverage data. The CRS stack can, thus, be seen as a generalization of conventional stacking velocity analysis as described in the previous section. The coherence may, again, be measured with the semblance measure, equation (3.5). The parameters yielding the highest coherence are also called *kinematic wavefield attributes*.

Using entire stacking surfaces instead of stacking trajectories to simulate zero-offset sections has a number of advantages:

- The number of traces contributing to the stack for each zero-offset location is considerably increased, which results in an improved S/N ration compared to conventional stacking methods as described in Section 3.2. This has been demonstrated in a number of data examples published, among others, by Mann et al. (1999) or Trappe et al. (2001).

- In cases where the S/N ratio or the number of traces in a CMP gather is too low to reliably determine a stacking velocity for the CMP stack, taking more CMP locations into account and directly fitting an entire surface to reflection events in the data may still be successful. This requires all three parameters in (3.7) to be determined simultaneously.

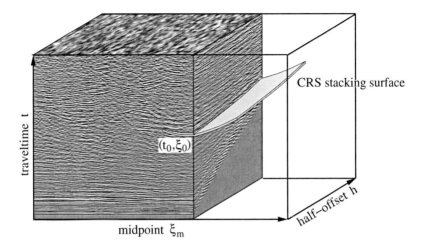

Figure 3.4: During the CRS stack, amplitudes in the multicoverage data are summed along optimum stacking surfaces in time–midpoint–half-offset ($t$-$\xi_m$-$h$) space. The stack result is assigned to the corresponding sample location $(t_0, \xi_0)$ in the zero-offset section.

- In the case of a laterally inhomogeneous subsurface with curved and dipping reflectors, the time-domain locations of recorded reflection amplitudes associated with a common-reflection point (CRP) are not confined to one midpoint location. The time-domain curve connecting these locations, known as *CRP trajectory*, is initially unknown. It can, however, be said that the second-order approximation of the CRP trajectory associated with a given zero-offset sample $(t_0, \xi_0)$ on a reflection event lies completely within the corresponding CRS stacking operator. It has been shown by Höcht et al. (1999) that once the kinematic wavefield attributes have been determined, and if the near-surface velocity is known, the second-order approximate CRP trajectory can be expressed through these quantities.

- The coefficients of the traveltime approximation (3.7), the kinematic wavefield attributes $p^{(\xi)}$, $M_N^{(\xi)}$, and $M_{\text{NIP}}^{(\xi)}$ determined during the CRS stack, contain information on the kinematics of the recorded wavefield (thus their name). As shown in Appendix A, they can be interpreted physically in terms of properties of two hypothetical wavefronts emerging at the measurement surface. These will be further discussed in Section 3.4. The kinematic wavefield attributes may be used for a number of different applications, including the calculation of geometrical spreading along the zero-offset ray (e. g., Hubral, 1983), or the determination of approximate projected Fresnel zones at the measurement surface (e. g., Mann, 2002). In particular, they contain information on the distribution of seismic velocities in the subsurface. This information will be used in the tomographic inversion method to be introduced in Chapter 4.

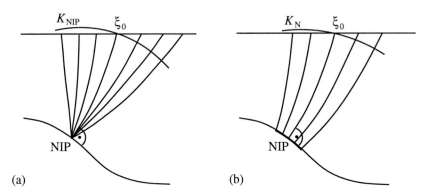

Figure 3.5: (a) NIP wave and (b) normal wave. The quantities $K_{\mathrm{NIP}}$ and $K_{\mathrm{N}}$ are the wavefront curvatures associated with the hypothetical emerging NIP and normal waves at $\xi_0$ on the measurement surface.

While the concept of the CRS stack method has here been introduced for the 2D case, it may also be applied in the case of 3D seismic data. As shown in Appendix A and further discussed in the following section, the 3D CRS operator depends on eight parameters. It should also be mentioned that the CRS stack method as described here is closely related to the multifocusing method introduced by Berkovitch et al. (1994) and Landa et al. (1999), where a different form of stacking operator is used.

## 3.4   Kinematic wavefield attributes

It is shown in Appendix A that the coefficients in the second-order traveltime approximation (3.7) used for the CRS stack can be related to kinematic properties of two hypothetical wavefronts emerging at the measurement surface location $\xi_0$. The quantities $p^{(\xi)}$ and $M_{\mathrm{N}}^{(\xi)}$ can be interpreted as the first and second horizontal spatial traveltime derivatives related to an emerging wavefront at $\xi_0$ due to an exploding reflector element placed at the normal-incidence point (NIP) of the zero-offset ray on the reflector (Figure 3.5b). As all rays associated with this wave are locally normal to the reflector element in the subsurface, it is known as the *normal wave*. The quantity $M_{\mathrm{NIP}}^{(\xi)}$ can be interpreted as the second horizontal traveltime derivative related to an emerging wavefront at $\xi_0$ due to a point source placed at the NIP on the reflector (Figure 3.5a). The associated wave is called the *NIP wave*. Identifying $M_{\mathrm{NIP}}^{(\xi)}$ with the NIP wave second spatial derivative requires the so-called NIP wave theorem (e. g., Chernyak and Gritsenko, 1979; Hubral, 1983) to be valid, which states that to second order in the offset coordinate, the CMP reflection traveltimes and the traveltimes along rays passing through the NIP of the zero-offset ray (Figure A.2) are identical.

If the near-surface velocity $v_0$ at $\xi_0$ is known and locally constant and subsurface structures are invariant in the direction perpendicular to the seismic line, the time-domain parameters $p^{(\xi)}$, $M_{\mathrm{N}}^{(\xi)}$, and $M_{NIP}^{(\xi)}$ in equation (3.7) can be related to quantities directly describing the emerging normal and NIP wavefronts in the vertical plane through the seismic line (see Appendix A):

$$p^{(\xi)} = \frac{\sin\alpha}{v_0} \, ,$$
$$M_N^{(\xi)} = \frac{\cos^2\alpha}{v_0} K_N \, , \qquad (3.8)$$
$$M_{NIP}^{(\xi)} = \frac{\cos^2\alpha}{v_0} K_{NIP} \, .$$

Here, $\alpha$ is the emergence angle (relative to the measurement surface normal) of the normal ray at $\xi_0$, while $K_N$ is the wavefront curvature of the emerging normal wave and $K_{NIP}$ is the wavefront curvature of the emerging NIP wave at $\xi_0$. Inserting expressions (3.8) in equation (3.7) yields the 2D CRS operator in the form

$$t^2(\xi_m, h) = \left(t_0 + \frac{2\sin\alpha}{v_0}\Delta\xi\right)^2 + \frac{2t_0\cos^2\alpha}{v_0}\left(K_N\Delta\xi^2 + K_{NIP}h^2\right) , \qquad (3.9)$$

which is identical to the equation published by Mann et al. (1999) and Jäger et al. (2001) if $K_N = R_N^{-1}$ and $K_{NIP} = R_{NIP}^{-1}$ are used. Here, $R_N$ and $R_{NIP}$ are the radii of wavefront curvature associated with the normal and NIP wave, respectively. The quantities $R_N$, $R_{NIP}$, and $\alpha$ are also sometimes called kinematic wavefield attributes. The term will here be used for any complete set of parameters determining the shape of the CRS operator.

By restricting equation (3.9) to $\xi_m = \xi_0$, that is, to $\Delta\xi = 0$ and comparing with (3.4), the kinematic wavefield attributes $R_{NIP}$ and $\alpha$ can be related to the normal-moveout velocity $v_{NMO}$:

$$v_{NMO}^2 = \frac{2v_0 R_{NIP}}{t_0\cos^2\alpha} \, . \qquad (3.10)$$

In the 3D case, the CRS operator can be written as (e. g., Schleicher et al., 1993; Höcht, 2002)

$$t^2(\xi_0 + \Delta\xi, h) = \left(t_0 + 2p^{(\xi)}\Delta\xi\right)^2 + 2t_0\left(\Delta\xi^T\underline{M}_N^{(\xi)}\Delta\xi + h^T\underline{M}_{NIP}^{(\xi)}h\right) , \qquad (3.11)$$

where the midpoint and half-offset coordinate vectors $\xi_m$ and $h$ are defined as in equation (3.1) and $\Delta\xi = \xi_m - \xi_0$. The quantity $p^{(\xi)}$ is a two-component vector, and $\underline{M}_N^{(\xi)}$ and $\underline{M}_{NIP}^{(\xi)}$ are symmetric $2 \times 2$ matrices. The 3D CRS operator, thus, depends on eight independent parameters.

In a way completely analogous to the 2D case, $p^{(\xi)}$, $\underline{M}_N^{(\xi)}$, and $\underline{M}_{NIP}^{(\xi)}$ can be related to kinematic properties of a normal wave and a NIP wave (Appendix A). The vector $p^{(\xi)}$ contains the first horizontal traveltime derivatives of the emerging normal and NIP waves at $\xi_0$. Its components are the horizontal components of the corresponding slowness vector. The matrix $\underline{M}_N^{(\xi)}$ contains the second traveltime derivatives of an emerging normal wave with respect to the spatial coordinates on the measurement surface at $\xi_0$, and $\underline{M}_{NIP}^{(\xi)}$ is the corresponding matrix of second spatial traveltime derivatives of an emerging NIP wave.

Again, if the near-surface velocity $v_0$ at $\xi_0$ is known and locally constant, the normal ray emergence direction can be determined from $\mathbf{p}^{(\xi)}$ and the matrices $\underline{\mathbf{M}}_N^{(\xi)}$ and $\underline{\mathbf{M}}_{NIP}^{(\xi)}$ can be related to wavefront curvatures of the normal and NIP wave. If $\alpha$ and $\psi$ denote the emergence angle and azimuth of the normal ray (Figure A.1), the following relations hold (see Appendix A):

$$\mathbf{p}^{(\xi)} = \frac{1}{v_0}(\sin\alpha\cos\psi, \sin\alpha\sin\psi)^T ,$$

$$\underline{\mathbf{M}}_N^{(\xi)} = \frac{1}{v_0}\mathbf{H}\mathbf{K}_N\mathbf{H}^T , \qquad (3.12)$$

$$\underline{\mathbf{M}}_{NIP}^{(\xi)} = \frac{1}{v_0}\mathbf{H}\mathbf{K}_{NIP}\mathbf{H}^T ,$$

where $\mathbf{K}_N$ and $\mathbf{K}_{NIP}$ are symmetric $2 \times 2$ matrices of wavefront curvature of the normal and NIP wave, respectively. The matrix $\mathbf{H}$ is the $2 \times 2$ upper left sub-matrix of the $3 \times 3$ transformation matrix from the local ray-centered Cartesian coordinate system to the global Cartesian coordinate system associated with the measurement surface, see Section 2.7 and Appendix A. Matrix $\mathbf{H}$ depends on $\alpha$ and $\psi$. The 3D CRS operator in terms of wavefront curvatures has also been derived by Höcht (2002). A similar expression can be found in Ursin (1982) and Gjøystdal et al. (1984).

The matrix $\underline{\mathbf{M}}_{NIP}^{(\xi)}$ may be related to the azimuth-dependent normal-moveout velocity $v_{NMO}(\phi)$ by

$$v_{NMO}^{-2}(\phi) = \frac{t_0}{2}\hat{\mathbf{e}}_\phi^T\underline{\mathbf{M}}_{NIP}^{(\xi)}\hat{\mathbf{e}}_\phi , \qquad (3.13)$$

where the two-component unit vector $\hat{\mathbf{e}}_\phi = (\cos\phi, \sin\phi)^T$ defines the azimuth (e. g., Hubral and Krey, 1980; Gjøystdal et al., 1984). If the seismic acquisition geometry is restricted to a certain azimuth range near a given direction $\phi$, as is usually the case in marine seismic acquisition, $\underline{\mathbf{M}}_{NIP}^{(\xi)}$ cannot be completely determined from the seismic data. Instead, only the component

$$M_\phi^{(\xi)} = \hat{\mathbf{e}}_\phi^T\underline{\mathbf{M}}_{NIP}^{(\xi)}\hat{\mathbf{e}}_\phi \qquad (3.14)$$

associated with the azimuth direction $\phi$ may then be obtained.

## 3.5 Practical aspects

In this section, practical aspects of the application of the common-reflection-surface stack method are discussed with special emphasis on the stable determination of the kinematic wavefield attributes for use in subsequent applications. Stable and reliable attributes are a prerequisite for the tomographic inversion process described in Chapter 4.

### 3.5.1 Results of the CRS stack process

During the CRS stack process, an optimum stacking operator is determined for each considered zero-offset sample $(t_0, \xi_0)$ in the 2D case, or $(t_0, \xi_0)$ in the 3D case. Thus, a set of kinematic wavefield attributes can be assigned to each zero-offset sample, resulting in a number of additional sections/volumes containing these attributes, together with the stacked zero-offset section/volume, itself. The kinematic wavefield attributes have meaningful values only where a reflection event is present in the multicoverage data. Only there can the attributes be interpreted physically in terms of normal and NIP wave parameters. Such locations are characterized by a high coherence along the CRS operator. An additional section/volume containing these coherence values therefore provides information on where reflection events could be detected and how reliable the associated kinematic wavefield attributes are. Note that the obtainable coherence values along a reflection event also depend on the S/N ratio for that event, on the number of contributing traces, and on the suitability of second-order traveltime approximations of the form of equations (3.7) or (3.11), respectively, to describe its shape.

The final result of the CRS stack process therefore consists of the stacked simulated zero-offset section/volume, a number of sections/volumes containing the kinematic wavefield attributes (three in the 2D case and up to eight in the 3D case), and a coherence section. Examples of these CRS stack results for a 2D real dataset are displayed in Figure 3.6.

### 3.5.2 Search strategies

Searching for optimum values for all kinematic wavefield attributes associated with a zero-offset sample simultaneously is a time-consuming process. Especially in the 3D case, where optimum values of up to eight parameters need to be found, efficient search strategies are essential. If the data quality and acquisition geometry allows, the search for the kinematic wavefield attributes may be split into a number of separate searches for a single parameter or a group of parameters in subsets of the multicoverage data.

One such strategy has been proposed by Mann et al. (1999) for the 2D case. It consists of first performing an automatic CMP stack by restricting equation (3.7) or (3.9) to $\Delta \xi = 0$ and performing a sample-by-sample stacking velocity analysis in each CMP gather, yielding $v_{\text{NMO}}$, which can be expressed through $R_{\text{NIP}}$ and $\alpha$, see equation (3.10). The resulting CMP stacked section can then be used to perform a search for $\alpha$ (or $p^{(\xi)}$ if the CRS operator in the form of equation (3.7) is used) with an operator obtained by setting $h = 0$ and $R_{\text{N}} = \infty$ (or $M_{\text{N}}^{(\xi)} = 0$, respectively) in the CRS operator. In a next step $R_{\text{N}}$ (or $M_{\text{N}}^{(\xi)}$) is determined using the CRS operator restricted to $h = 0$. Finally, the obtained values may be used as the starting point for a local optimization of the kinematic wavefield attribute values in the full multicoverage data. Prerequisite for the application of this strategy is that each parameter can be determined in a stable way in the respective subset of the multicoverage data. In particular, there needs to be a sufficient number of traces in each data subset and the S/N ratio needs to be sufficiently high.

If conflicting dips are considered, that is, more than one value for $\alpha$ (or $p^{(\xi)}$) is allowed at each zero-offset sample, a different strategy needs to be applied (e. g., Mann, 2002). In the case of 3D seismic data, the appropriate search strategy to be used strongly depends on the acquisition geometry of the data.

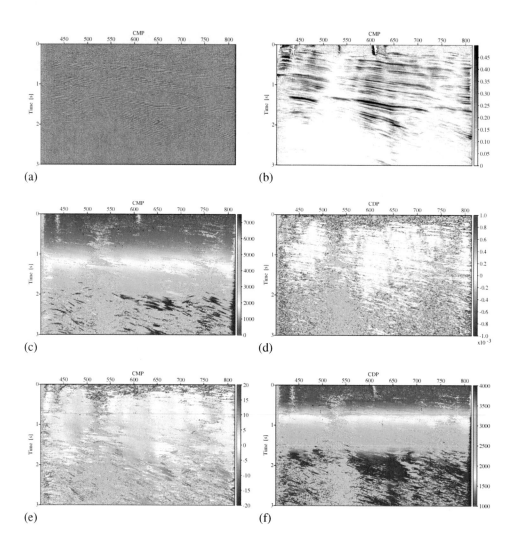

Figure 3.6: Results of the CRS stack process for a 2D real data example (compare Section 6.3). These include: (a) Simulated zero-offset section, (b) CRS coherence section (semblance), (c) $R_{NIP}$ section (displayed values are in m), (d) $K_N = R_N^{-1}$ section (displayed values are in m$^{-1}$), (e) emergence angle $\alpha$ section (displayed values are in degrees), (f) stacking velocity section (displayed values are in m/s). In Figures (c) to (f), attributes corresponding to low coherence values are masked. In addition to the displayed sections, the intermediate results of the one-parameter searches (if applied) and an automatic time-migrated section are obtained.

### 3.5.3 Aperture considerations

The validity of the approximations (3.7) and (3.11) for the description of reflection traveltimes generally decreases with increasing distance in the midpoint and offset directions from the considered zero-offset location. The aperture (in midpoint and offset) used during the parameter search and stacking processes of the CRS stack need to be chosen appropriately to obtain optimum results.

If the main objective is the extraction of kinematic wavefield attributes from the data, special care needs to be taken to control the effects of spread-length-bias (Section 3.2). While for large apertures, the second-order approximation may no longer be valid and the optimum operator may fit to a reflection event not at the considered zero-offset location, but at larger offsets, a small aperture will decrease the resolution with which the attributes can be determined. This effect is well known from conventional stacking velocity analysis.

The proper choice of the offset and midpoint apertures depends on the characteristics of the dataset under consideration. If the traveltime moveout is far from hyperbolic due to a very complex subsurface, reliable attributes may in some cases not be obtainable at all. The effect of the aperture size and other parameters on the reliability of kinematic wavefield attributes is further discussed in Section 6.1 in the context of the attribute-based tomographic inversion.

### 3.5.4 Smoothing of attributes

In conventional stacking velocity analysis, stacking velocity values are usually picked at a number of selected zero-offset traveltimes in the velocity spectrum and are interpolated in between. This leads to an offset-dependent distortion of the wavelet shape, especially for shallow events, known as NMO stretch (e. g., Yilmaz, 2001). Due to the fact that during the CRS stack, the optimum stacking operator is determined independently for each sample in the zero-offset section/volume to be simulated, this NMO stretch effect is avoided (Mann and Höcht, 2003).

The separate sample-by-sample determination of the stacking parameters may, however, lead to unwanted fluctuations of attributes in the obtained kinematic wavefield attribute sections, as a stable determination of attributes may not be possible at every zero-offset sample location. These fluctuations may have adverse effects not only on the stack result itself, but also on the use of the kinematic wavefield attributes for further processes like velocity model estimation.

On the other hand, in contrast to $v_{NMO}$, moveout parameters that are directly linked to spatial traveltime derivatives remain locally constant along the time axis on the wavelet. This is also true for the kinematic wavefield attributes discussed in Section 3.4. In addition, as long as the assumptions made during the CRS stack are valid (applicability of paraxial ray theory), attributes that are proportional to traveltime derivatives should vary smoothly along an event in the spatial direction. These two observations justify the application of a smoothing process to kinematic wavefield attribute sections/volumes prior to performing the final stack or using the attributes for other purposes. A simple event-consistent smoothing algorithm for kinematic wavefield attribute sections is described in Appendix G. Such a smoothing of attribute sections/volumes can significantly enhance the stack result and may allow to omit the time-consuming local optimization of attributes mentioned above.

# Chapter 4

# Tomographic inversion with kinematic wavefield attributes

The CRS stack process described in the previous chapter can be regarded as a tool for automatically extracting traveltime information from the seismic data in the form of kinematic wavefield attributes. If the traveltimes of reflection events in the data are reasonably well described by the second-order approximations (3.7) or (3.11), respectively, the information contained in the kinematic wavefield attributes can be used for the determination of a laterally inhomogeneous velocity model for depth imaging.

In this chapter, a tomographic inversion method based on kinematic wavefield attributes will be introduced. While the use of attributes, and thus of traveltime approximations, for velocity model estimation limits the allowed degree of complexity of the subsurface velocity structure, it leads to a number of clear practical advantages, particularly in the case of seismic data with a low S/N ratio. In such data it may be difficult, or even impossible, to reliably identify and pick reflection events in the prestack data, as is required for conventional reflection tomography. Traveltime approximations like equation (3.7) or (3.11), on the other hand, allow to automatically correlate reflection events on a large number of traces with varying midpoints and offsets, making it possible to identify reflections and determine their traveltimes even when the S/N ratio is low. Also, because the offset-dependence of traveltimes is already contained in the determined kinematic wavefield attributes, the amount of picking required to obtain the input for the tomographic inversion is significantly reduced. Picking can be performed directly in the stacked simulated zero-offset section/volume of improved S/N ratio obtained with the CRS stack.

## 4.1 Smooth versus layered velocity models

An important aspect to consider when determining a velocity model for depth migration is the parametrization of the model. As discussed by Jannane et al. (1989) and Claerbout (1985), the seismic data recorded at the Earth's surface do not contain sufficient information to resolve the true subsurface velocity distribution at all length scales. There exists an information gap leading to an ambiguity, or null space, and all one can hope to determine is a velocity model that is consistent with the data. In order to ensure a unique and stable solution of the inversion problem, it is, therefore, required to make assumptions about the subsurface velocity distribution. Constraints may for example be introduced by requiring velocities to be smooth, to vary according to simple velocity laws that can be represented with a limited number of parameters, or to have discontinuities only at a number of predefined interfaces. The different kinds of constraints are reflected in the different types of model parametrizations usually used in seismic velocity estimation. These fall into three main categories (see Figure 4.1):

- Layered or blocky velocity models, where the velocity in each block or layer is either constant or may vary according to a simple velocity law (vertical or horizontal velocity gradients). The velocity may be discontinuous at block or layer boundaries.

- Gridded or smooth velocity models which do not contain any velocity discontinuities. The velocity is either defined on a (dense) grid of subsurface points, varying smoothly from gridpoint to gridpoint, or it is defined analytically everywhere in the model using smooth functions.

- Hybrid models consisting of a smooth or gridded background model that contains irregularly shaped bodies of high velocity contrast.

For the proper choice of a velocity model type, several factors need to be taken into account. First of all, the type of model should be suitable for the investigated geological environment. If sedimentary layers with velocity contrasts from layer to layer are expected, a layered or blocky velocity model may be a suitable choice, while hybrid models are more appropriate in the presence of high-contrast salt bodies. Secondly, the choice of velocity model type is directly related to the model estimation method that is used to determine the model parameters. Each velocity estimation method usually implicitly assumes a certain type of model description. For example, methods that use layer stripping require layer boundaries to be defined in the model. In the following, geological situations with only moderate lateral contrasts in velocity will be assumed. In such situations, it is in principle possible to use either smooth or blocky velocity models.

An important point to be considered for the choice of a suitable model type is the information that is available in the seismic data. The drawback of blocky or layered models is that the continuous model interfaces have to be associated with reflection events in the seismic data. It is, however, often difficult to identify reflection events that are continuous across large regions of the seismic section or volume, either because of the complexity of subsurface reflector structures or because of regions of low S/N ratio in the seismic data, where clear reflection events are not visible at all.

In smooth models without discontinuities, on the other hand, subsurface reflector locations and the velocity distribution can be treated independently. The smooth velocity model itself then represents the long-wavelength component of the subsurface velocity distribution. This separate treatment of subsurface reflectors and a smooth velocity distribution is in accordance with the assumptions made in seismic imaging methods based on the Born-approximation (e. g., Bleistein

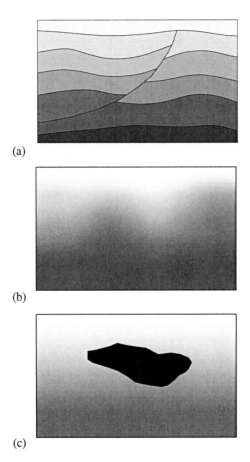

(a)

(b)

(c)

Figure 4.1: Different types of velocity models commonly used for depth imaging. (a) Blocky velocity model, (b) smooth or gridded velocity model, (c) hybrid velocity model. See main text for details.

et al., 2001). It is justified by the well-known fact that the kinematics of seismic wave propagation are mainly controlled by the long-wavelength component of velocity, while reflection amplitudes can be associated with discontinuities in the velocity, or the short wavelength component. It is impossible to determine short-wavelength features of the subsurface distribution of seismic velocities from measured traveltimes of seismic waves alone (e. g., Jannane et al., 1989). This also means that for the purposes of structural seismic imaging, a smoothed version of the true velocity distribution is usually sufficient. In the context of velocity model determination for depth imaging, it is, thus, reasonable to use a smooth model description. The influence of the degree of smoothing of velocities on the result of depth migration has been investigated by Versteeg (1993).

The use of smooth velocity models for traveltime inversion (see below) allows reflection points in the subsurface corresponding to reflection events in the seismic data to be considered independently of each other. There is no need to assume continuous reflectors in the model. It, thus, becomes possible to formulate inversion algorithms that require only locally coherent reflection events in the data. Further advantages of smooth models as opposed to layered or blocky models in the context of traveltime inversion/reflection tomography are discussed in Lailly and Sinoquet (1996).

## 4.2 NIP waves and velocities model estimation

The criterion of consistency of velocity models with the seismic data used in migration-based velocity analysis as described in Section 1.3 leads to the following statement: a model in which all reflection signals in the seismic data pertaining to a common reflection point (CRP) in the true subsurface are migrated to a common point is consistent with the data.

The ray segments of the corresponding specular rays connecting sources and receivers on the measurement surface with a CRP in the model (Figure 4.2a) are geometrically identical to ray trajectories associated with a hypothetical emerging wave due to a point source at the CRP (Figure 4.2b). In the previous chapter, such a wave has been called NIP wave, as the considered CRP is identical to the normal-incidence point (NIP) of the zero-offset ray on the reflector. If reflection traveltimes in the seismic data pertaining to a CRP can be interpreted in terms of NIP wave traveltimes, the imaging of the associated reflection signals to a common point in the model is equivalent to the focusing of the NIP wave at zero traveltime at that point (Figure 4.2c). Thus, a model in which all NIP waves, when propagated back into the subsurface, focus at zero traveltime is consistent with the data. This criterion for a consistent model is in accordance with the criterion of depth-focusing analysis as described above: a migration velocity model is consistent if seismic reflections, when downward continuation is performed, focus at zero traveltime.

It has been shown in Chapter 3 that the parameters describing a second-order approximation of the traveltimes of emerging NIP wavefronts can be extracted from the seismic multicoverage data by applying the CRS stack or by performing conventional stacking velocity analysis and an additional local zero-offset dip search. In Chapter 3 the involved parameters have been called kinematic wavefield attributes. They describe the emerging hypothetical NIP wavefront either in terms of first and second traveltime derivatives or—if the near-surface velocity is known—in terms of its

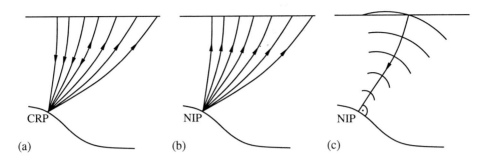

Figure 4.2: (a) The ray segments of specular rays reflected at a CRP in the subsurface. (b) The ray trajectories associated with a hypothetical wave due to a point source at the NIP (NIP wave). Geometrically, the CRP ray segments and the NIP wave ray segments coincide. (c) In a consistent velocity model, NIP waves focus at the NIP at zero traveltime, when they are propagated back into the subsurface.

emergence direction and wavefront curvature. Once the kinematic wavefield attributes associated with an emerging NIP wave and the near-surface velocity at the corresponding normal ray emergence location are known, a second-order approximation of the CRP trajectory (the curve in the prestack data along which signals pertaining to a CRP may be found) can be determined (Höcht et al., 1999).

Hubral and Krey (1980) make extensive use of the concept of "having a NIP wavefront shrink back to its hypothetical source" for the determination of interval velocities from stacking velocities and zero-offset dips. They present 2D and 3D Dix-type algorithms for the construction of models consisting of iso-velocity layers separated by curved interfaces, based on the focusing of NIP waves. Assuming the upper $n-1$ layer velocities and curved velocity interfaces to be known, these algorithms allow to construct the $n$th interface and layer velocity by propagating NIP waves pertaining to the $n$th reflector through the $n-1$ constant-velocity layers in the overburden using wavefront curvature transmission and refraction laws and requiring the radius of wavefront curvature to become zero on the $n$th interface. For the algorithm to work, stacking velocity and dip information from continuous reflection events in the data needs to be available in order to be able to construct the curved velocity interfaces. The NIP wave curvature is calculated from stacking velocities using equations (3.10) and (3.13). For the 3D inversion, the stacking velocity is required in only one azimuth direction. In the 1D case, the inversion concept of Hubral and Krey (1980) reduces to conventional Dix inversion, which can be formulated in terms of NIP wave radii of curvature by setting $\alpha = 0$ in equation (3.10). A similar 3D algorithm has been described by Chernyak and Gritsenko (1979). Biloti et al. (2002) present a 2D algorithm based on the focusing of NIP waves for the determination of layers with constant velocity gradients, separated by curved interfaces. Inversion algorithms based on wavefront curvature and emergence angle information obtained from multifocusing results have been presented by Berkovitch and Gelchinsky (1989) and Keydar et al. (1995).

The use of traveltime information in the form of kinematic wavefield attributes related to hypothetical emerging NIP waves—and, thus, of second-order traveltime approximations—in general limits the applicability of velocity estimation methods to velocity distributions of moderate lateral variation. However, the distinct advantage of using kinematic wavefield attributes lies in the fact that these attributes can be determined from the seismic prestack data even in the presence of a low overall S/N ratio, when identifying reflection events on prestack traces becomes difficult. Also, compared to traveltime inversion methods based on prestack traveltimes, the required picking effort is significantly reduced as picking can be performed in the stacked simulated zero-offset section/volume of increased S/N ratio.

The concept of determining velocity models by requiring NIP waves—described by kinematic wavefield attributes—to focus at zero traveltime can also be extended to the case of smooth, laterally inhomogeneous velocity distributions. As discussed above, the use of such models in velocity estimation methods leads to further advantages, as it allows the formulation of inversion algorithms which do not assume continuous reflectors in the model. Pick locations in the stacked zero-offset section/volume can then, in principle, be independent of each other, which further simplifies the picking process. In the following section/volume, a NIP-wave based tomographic inversion method for the determination of smooth velocity models will be introduced, that makes use of this advantage. Pick locations in the simulated zero-offset section/volume required for this approach do not need to follow continuous reflection events in the data, but may be located on events that are only locally coherent. In particular, the determination of velocity models becomes possible also for data in which it is difficult to follow reflection events continuously across the seismic section/volume, as long as the kinematic wavefield attributes can be expected to be reliably determined and the overall pick density across the seismic section/volume is sufficient.

## 4.3 Formulation of tomography with kinematic wavefield attributes

In this section, the kinematic wavefield attributes associated with hypothetical emerging NIP waves will be used to formulate a tomographic inversion method for the determination of smooth isotropic velocity models. The method will first be introduced for the most general case, the determination of 3D velocity models with the full set of kinematic wavefield attributes assumed to be available for each considered NIP wave. It will then be discussed in more detail for the special cases of 1D and 2D inversion and the case of 3D inversion with limited azimuth information in Chapter 5.

### 4.3.1 Data components

In the general 3D case, a hypothetical emerging NIP wave associated with a given zero-offset sample $(t_0, \boldsymbol{\xi}_0)$ on a reflection event is characterized by its normal ray traveltime $\tau_0 = t_0/2$, its first spatial traveltime derivatives (or horizontal slowness components) given by vector $\mathbf{p}^{(\xi)}$, and its second spatial traveltime derivatives given by the symmetric matrix $\underline{\mathbf{M}}_{\mathrm{NIP}}^{(\xi)}$ (see Section 3.4). An emerging NIP wave can, thus, be represented by

$$\left( \tau_0, \underline{\mathbf{M}}_{\mathrm{NIP}}^{(\xi)}, \mathbf{p}^{(\xi)}, \boldsymbol{\xi}_0 \right) , \tag{4.1}$$

which will be referred to as a data point. Such data points can be obtained directly from the CRS stack results by picking zero-offset samples $(t_0, \boldsymbol{\xi}_0)$ on reflection events in the CRS stacked volume and extracting the elements of $\underline{\mathbf{M}}^{(\xi)}_{\mathrm{NIP}}$ and $\mathbf{p}^{(\xi)}$ from the corresponding locations in the kinematic wavefield attribute volumes. Pick locations are independent of each other and do not need to follow reflection events in the zero-offset volume over successive traces. Picks may lie on reflection events that are only locally coherent.

A direct way of implementing the criterion of focusing of NIP waves for the determination of a consistent smooth velocity model would be to propagate the NIP wavefronts associated with the data points (4.1) into the subsurface and check if they focus at $\tau_0 = 0$. Focusing implies that the NIP wave radius of wavefront curvature, given by the matrix $\mathbf{R}_{\mathrm{NIP}} = \mathbf{K}^{-1}_{\mathrm{NIP}}$, becomes zero. One would then need to find a model that effects this focusing for all considered data points. The initial propagation direction of a NIP wave corresponding to a data point (4.1) is determined by its horizontal slowness vector $\mathbf{p}^{(\xi)}$ at the location $\boldsymbol{\xi}_0$ and the local near-surface velocity value in the given model. The subsurface location at which the focusing criterion is evaluated is defined by the normal ray traveltime $\tau_0$. All data components must, however, be expected to be affected by noise or a certain measurement error: the components of $\underline{\mathbf{M}}^{(\xi)}_{\mathrm{NIP}}$ and $\mathbf{p}^{(\xi)}$ have been determined with a coherence analysis from a finite number of traces in the (possibly noisy) seismic data, while $\tau_0$ has been picked on a reflection signal of finite length on a seismic trace and may carry a picking error, thus not representing the true, exact reflection traveltime. Fixing $\tau_0$ and $\mathbf{p}^{(\xi)}$ would mean ignoring possible measurement errors in these quantities, which may lead to a destabilization of the inversion process.

If, on the other hand, the propagation of NIP waves through a given model is started at the respective NIP in the subsurface, possible noise or measurement errors in all data components may be accounted for. A velocity model is then consistent with the data if all data components (4.1) of all considered NIP waves are correctly modeled in the sense that the misfit between the data components and the corresponding forward-modeled quantities is minimized and falls below a certain error threshold. The aim of the inversion procedure is then to find such an optimum model. This is the approach which will be followed here. It is similar to that used by Billette and Lambaré (1998) in the context of stereotomography.

### 4.3.2 Model components

The true subsurface locations of the NIPs corresponding to the data points (4.1) and the associated local reflector dips, defining the normal ray direction at the respective NIP, are initially unknown. If in the inversion process the propagation of the NIP waves through the velocity model is started in the subsurface, these quantities need to be considered as additional model parameters to be determined during the inversion together with the velocity distribution. In the general 3D case, a NIP in the subsurface is characterized by three spatial coordinates and two parameters defining the local reflector dip at the NIP. These may either be two components of a unit vector normal to the reflector at the NIP or two angles.

In this chapter, global Cartesian coordinates will be denoted by $(x, y, z)$. While this nomenclature deviates from that used in Chapter 2, it is more convenient in the present context. The positive

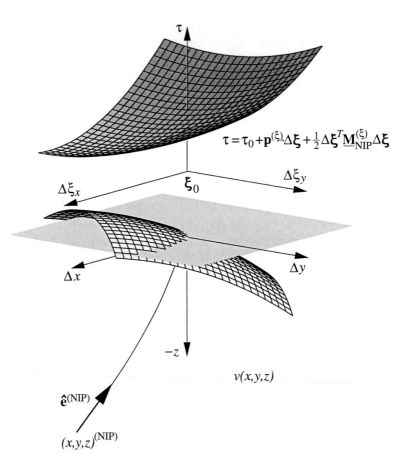

Figure 4.3: Definition of data and model components in the NIP-wave-based tomographic inversion. The components of each data point (4.1) describe the second-order traveltime surface associated with an emerging NIP wavefront. The corresponding NIP location and normal ray initial direction in the subsurface, equation (4.2), are considered as model parameters, together with the B-spline coefficients describing the smooth velocity distribution itself, equation (4.3). Here, $\Delta\boldsymbol{\xi}$ is defined as $\Delta\boldsymbol{\xi} = \boldsymbol{\xi} - \boldsymbol{\xi}_0$, where $\boldsymbol{\xi}_0$ is the emergence location of the considered normal ray. The horizontal coordinates $\Delta\xi_x$ and $\Delta\xi_y$ are identical to the relative coordinates $\Delta x$ and $\Delta y$.

$z$-direction is assumed to point upwards with the measurement surface at $z = 0$. Points on the measurement surface will be denoted by two-component vectors $\boldsymbol{\xi}$, in agreement with the notation in Chapter 3. With this notation, the quantities characterizing a NIP in the subsurface may be written as

$$(x, y, z, e_x, e_y)^{(\text{NIP})}, \tag{4.2}$$

where $x^{(\text{NIP})}$, $y^{(\text{NIP})}$, and $z^{(\text{NIP})}$ are the spatial coordinates of the NIP, while $e_x^{(\text{NIP})}$ and $e_y^{(\text{NIP})}$ are the horizontal components of a unit vector $\hat{\mathbf{e}}^{(\text{NIP})}$ that is locally normal to the reflector. Note that $z^{(\text{NIP})} < 0$, as all NIPs are located below the measurement surface $z = 0$. Each data point (4.1) representing an emerging NIP wave, is associated with a set of model parameters, equation (4.2), which represent the corresponding NIP in the subsurface.

The smooth velocity model itself is represented by B-splines (de Boor, 1978). If a three-dimensional grid is defined by two strictly increasing knot sequences in the $x$- and $y$-directions and a strictly increasing knot sequence in the negative $z$-direction, the velocity model can be written in terms of B-splines (Appendix B) as

$$v(x, y, z) = \sum_{j=1}^{n_x} \sum_{k=1}^{n_y} \sum_{l=1}^{n_z} v_{jkl} \, \beta_j(x) \, \beta_k(y) \, \beta_l(-z), \tag{4.3}$$

where $\beta_j(x)$, $\beta_k(y)$, and $\beta_l(-z)$ are B-spline basis functions of a given degree $m$, and $v_{jkl}$ are the B-spline coefficients. The minus sign in equation (4.3) has been introduced to account for the fact that the positive $z$-direction points upwards, while the corresponding B-spline knot sequence is defined to be increasing with depth, see Appendix B.

B-spline basis functions are spline functions of minimum length on a given knot sequence, that is, they are non-zero only within a range of $m + 1$ knot intervals. Being defined as a linear combination of splines of degree $m$, the function $v(x, y, z)$ in equation (4.3) has continuous $(m - 1)$st derivatives. Thus, B-splines provide an

- analytical, smooth model description with

- continuous derivatives up to $(m - 1)$st order with a

- limited number of model parameters associated with

- localized basis functions.

These properties make B-splines ideal for tomographic inversion applications. The smoothness properties of the model are required to allow the application of ray-tracing methods based on the equations given in Chapter 2, while a model description with as few parameters as possible and with local control is desirable to allow an efficient and unique solution of the inversion problem. For the 2D and 3D case, the tomographic method described here requires continuous third derivatives of the velocity field. Thus, B-splines of degree $m = 4$ will be used in these cases, while for 1D tomographic inversion, the velocity model will be described in terms of cubic B-splines ($m = 3$).

### 4.3.3 Forward modeling

In a given velocity model, the kinematic wavefield attributes of NIP waves associated with given NIP model parameters (4.2) can be calculated by forward modeling. For that purpose, it is sufficient to trace the normal ray starting at the respective NIP with kinematic ray tracing and perform dynamic ray tracing along that ray. While the two-component vector $\mathbf{p}^{(\xi)}$ simply contains the horizontal components of the three-component slowness vector of the normal ray emerging at $\boldsymbol{\xi}_0$, the normal ray traveltime $\tau_0$ is obtained by integration of a suitable form of equation (2.17) along the ray. The symmetric $2 \times 2$ matrix $\underline{\mathbf{M}}_{\mathrm{NIP}}^{(\xi)}$ can be calculated from the elements of the ray-propagator matrix computed along the normal ray by integrating the system of equations (2.26) adapted to the chosen coordinate system. The required quantities are the solutions of the dynamic ray-tracing system (2.34) for point source initial conditions.

If ray-centered coordinates (Section 2.5) are used and the near-surface velocity at $\boldsymbol{\xi}_0$ is assumed to be locally constant, then

$$\underline{\mathbf{M}}_{\mathrm{NIP}}^{(\xi)} = \underline{\mathbf{H}}\, \underline{\mathbf{M}}_{\mathrm{NIP}}\, \underline{\mathbf{H}}^T \quad \text{with} \quad \underline{\mathbf{M}}_{\mathrm{NIP}} = \underline{\mathbf{P}}_2\, \underline{\mathbf{Q}}_2^{-1} \,. \tag{4.4}$$

Here, as in Section 2.7, $\underline{\mathbf{H}}$ is the upper left $2 \times 2$ submatrix of the $3 \times 3$ transformation matrix from local ray-centered Cartesian coordinates at the ray endpoint to the global Cartesian coordinate system, in this case attached to the measurement surface. If the $q_1$-direction of the ray-centered coordinate system lies in the vertical plane spanned by the slowness vector and the normal to the measurement surface, $\underline{\mathbf{H}}$ is given by expression (A.6).

If dynamic ray-tracing is performed in global Cartesian coordinates using the reduced-Hamiltonian formulation of Section 2.6 with the $z$-coordinate as the independent parameter along the ray, matrix $\underline{\mathbf{M}}_{\mathrm{NIP}}^{(\xi)}$ is directly given by expression (2.85):

$$\underline{\mathbf{M}}_{\mathrm{NIP}}^{(\xi)} = \underline{\mathbf{P}}_2^{(x)}\, \underline{\mathbf{Q}}_2^{(x)\,-1} \,. \tag{4.5}$$

### 4.3.4 Inverse problem

With the model and data components defined above, the inverse problem to be solved can be formulated for the general 3D case as follows:

Given a total of $n_{\mathrm{data}}$ data points

$$\left( \tau_0, \underline{\mathbf{M}}_{\mathrm{NIP}}^{(\xi)}, \mathbf{p}^{(\xi)}, \boldsymbol{\xi}_0 \right)_i \qquad i = 1, \ldots, n_{\mathrm{data}} \,, \tag{4.6}$$

picked from the CRS stack results (or obtained in another way from the seismic data), find a model defined by parameters

$$\begin{aligned} (x, y, z, e_x, e_y)_i^{(\mathrm{NIP})} \qquad & i = 1, \ldots, n_{\mathrm{data}} \,, \\ v_{jkl} \qquad & j = 1, \ldots, n_x \,, \quad k = 1, \ldots, n_y \,, \quad l = 1, \ldots, n_z \,, \end{aligned} \tag{4.7}$$

such that the misfit between the picked data points (4.6) and the corresponding forward-modeled quantities

$$(\tau_0, \mathbf{M}_{\mathrm{NIP}}^{(\xi)}, \mathbf{p}^{(\xi)}, \boldsymbol{\xi}_0)_i^{\mathrm{mod}} \qquad i = 1, \ldots, n_{\mathrm{data}} , \tag{4.8}$$

associated with the NIP model parameters given in (4.7), is minimized. A procedure for solving this inverse problem will be presented in the following section. For that purpose, the inverse problem will be formulated as a least-squares problem.

## 4.4 Solution of the inverse problem

If the data components (4.6) and the model components (4.7) are rearranged into a data vector $\mathbf{d}$ and a model vector $\mathbf{m}$, the inverse problem formulated in the previous section can be restated as follows: an optimum model $\mathbf{m}$ is sought, such that the misfit between the data $\mathbf{d}$ and the corresponding modeled values $\mathbf{d}_{\mathrm{mod}} = \mathbf{f}(\mathbf{m})$ is minimized. The nonlinear operator $\mathbf{f}$ symbolizes the forward modeling by dynamic ray tracing to obtain the quantities (4.8). If the least-squares norm (Tarantola, 1987), or weighted $L_2$ norm, is used as a measure of misfit, the optimum model is found by minimizing a cost function

$$S(\mathbf{m}) = \frac{1}{2}\|\mathbf{d} - \mathbf{f}(\mathbf{m})\|_D^2 = \frac{1}{2}\|\Delta\mathbf{d}(\mathbf{m})\|_D^2 = \frac{1}{2}\Delta\mathbf{d}^T(\mathbf{m})\underline{\mathbf{C}}_D^{-1}\Delta\mathbf{d}(\mathbf{m}) , \tag{4.9}$$

where $\Delta\mathbf{d}(\mathbf{m}) = \mathbf{d} - \mathbf{f}(\mathbf{m})$ and $\underline{\mathbf{C}}_D$ is a symmetric and positive definite matrix whose elements act as weights applied to the different data components in the calculation of $S$.

The matrix $\underline{\mathbf{C}}_D$ is sometimes interpreted as a data covariance matrix (Tarantola, 1987). Its diagonal elements $(C_D)_{ii} = \sigma_i^2$ are then the variances (squared standard deviations) associated with the different data components. Thus, each data point is weighted according to its uncertainty or expected data error. Here, $\underline{\mathbf{C}}_D$ is assumed to be diagonal, implying that data errors are uncorrelated. In practice, weighting of the different data components is also required to account for the fact that different types of data with different physical dimensions are involved. Weights, or scaling factors, are required to stabilize the inversion process by bringing the numerical values of the different types of data to a comparable size. Each data point given by equation (4.1) contains four different data types: traveltime, second traveltime derivative, first traveltime derivative, and spatial coordinate. Suitable choices for the corresponding scaling factors $\sigma_\tau$, $\sigma_M$, $\sigma_p$, and $\sigma_\xi$ are discussed in Chapter 5.

Due to the nonlinearity of the forward modeling operator $\mathbf{f}$ (the nonlinear dependence of dynamic ray-tracing results on the model parameters), the inverse problem to be solved by minimizing the cost function (4.9) is also nonlinear. In principle, its solution requires the application of global nonlinear optimization methods (e.g., Sen and Stoffa, 1995). Such methods are, however, computationally very expensive. Therefore, an iterative, local approach to solving the inverse problem will be used here.

In the vicinity of a given model vector $\mathbf{m}_n$ corresponding to the $n$th iteration, the modeling operator can be locally approximated by

$$\mathbf{f}(\mathbf{m}_n + \Delta\mathbf{m}) \approx \mathbf{f}(\mathbf{m}_n) + \underline{\mathbf{F}}\Delta\mathbf{m} , \tag{4.10}$$

where $\underline{\mathbf{F}}$ is a matrix that contains the derivatives $F_{ij} = \partial f_i / \partial m_j$, also known as *Fréchet derivatives*, of $\mathbf{f}$ at $\mathbf{m}_n$. Their calculation will be discussed below and in Appendices D and E. The linear approximation (4.10) of the modeling operator allows a minimum of $S$ to be found iteratively by computing the least-squares solution to the locally linearized problem during each iteration step. Starting with a first-guess model $\mathbf{m}_0$, a sequence of model updates $\Delta\mathbf{m}$ is found which, under favorable conditions, leads to convergence to the global minimum of $S$.

### 4.4.1   Least-squares solution

A necessary condition for the cost function $S$, equation (4.9), to have a minimum is the vanishing of its first derivatives with respect to the model parameters,

$$\nabla_{\mathbf{m}} S = \mathbf{0} \ . \tag{4.11}$$

Taking the gradient of $S$ with respect to $\mathbf{m}$ yields

$$
\begin{aligned}
\nabla_{\mathbf{m}} S &= -\underline{\mathbf{F}}^T \underline{\mathbf{C}}_D^{-1} \Delta\mathbf{d}(\mathbf{m}) \\
&= -\underline{\mathbf{F}}^T \underline{\mathbf{C}}_D^{-1} (\mathbf{d} - \mathbf{f}(\mathbf{m})) \\
&= -\underline{\mathbf{F}}^T \underline{\mathbf{C}}_D^{-1} (\Delta\mathbf{d}(\mathbf{m}_n) - \underline{\mathbf{F}}\Delta\mathbf{m}) \ ,
\end{aligned}
\tag{4.12}
$$

where equation (4.10) has been used, implying that $\nabla_{\mathbf{m}} S$ is assumed to be evaluated near $\mathbf{m}_n$. Setting $\nabla_{\mathbf{m}} S$ equal to zero results in

$$\underline{\mathbf{F}}^T \underline{\mathbf{C}}_D^{-1} \underline{\mathbf{F}}\Delta\mathbf{m} = \underline{\mathbf{F}}^T \underline{\mathbf{C}}_D^{-1} \Delta\mathbf{d}(\mathbf{m}_n) \ , \tag{4.13}$$

which, except for the additional matrix $\underline{\mathbf{C}}_D^{-1}$, is identical to the well-known normal equations (e. g., Lines and Treitel, 1984). Equation (4.13) yields the least-squares solution to the linear system

$$\underline{\mathbf{C}}_D^{-1/2} \underline{\mathbf{F}}\Delta\mathbf{m} = \underline{\mathbf{C}}_D^{-1/2} \Delta\mathbf{d}(\mathbf{m}_n) \ . \tag{4.14}$$

It can be solved for $\Delta\mathbf{m}$ if the inverse of $\underline{\mathbf{F}}^T \underline{\mathbf{C}}_D^{-1} \underline{\mathbf{F}}$ exists and can be computed in a stable way. Methods for computing the least-squares solution of a linear system of equations without having to explicitly compute the transpose or inverse of a matrix are discussed in Sections 5.1 and 5.2.

Problems occur when $\underline{\mathbf{F}}^T \underline{\mathbf{C}}_D^{-1} \underline{\mathbf{F}}$ is singular or near-singular and a stable inverse cannot be computed. This means that the system of equations (4.14) does not contain sufficient information to uniquely determine all model parameters. This situation is frequently encountered in tomographic problems, including the one discussed here. In such cases, additional constraints need to be introduced to regularize the problem.

### 4.4.2   Regularization

A common way of introducing additional constraints on the model for regularization is to require the model (or model update) vector to have minimum length. This is accomplished by adding an extra term to the cost function, consisting of the squared $L_2$ norm of the model vector multiplied by a positive weighting or damping factor. Adding this extra term to the cost function effectively amounts to adding the corresponding damping factor to all diagonal elements of matrix $\mathbf{F}^T\underline{\mathbf{C}}_D^{-1}\mathbf{F}$, thus ensuring its invertability (Menke, 1984). This method is known as the Marquardt-Levenberg method (Levenberg, 1944; Marquardt, 1963), or damped least-squares.

However, in the context of the tomographic inversion problem discussed here, a physically more sensible way of constraining the model parameters is to require the velocity model to have minimum second derivatives. The second derivatives give a measure of curvature and, thus, of roughness of the velocity model. Minimizing them is a reasonable constraint, as the simplest or smoothest model that explains the data is sought during the inversion. The smoothness requirement also ensures a sufficient range of validity of paraxial ray theory around each considered central ray, allowing to relate the computed kinematic wavefield attributes to those obtained from the seismic data.

Therefore, instead of the Marquardt-Levenberg constraint of minimum $L_2$ norm of the model vector, the minimum curvature constraint will be used here. It is shown in Appendix C how this constraint, applied to the velocity model $v(x,y,z)$, can be realized by adding an extra term involving only the velocity model parameters $v_{jkl}$ to the cost function. If that part of the model vector which contains these velocity model parameters is denoted by $\mathbf{m}^{(v)}$, such that

$$\mathbf{m} = \begin{pmatrix} \mathbf{m}^{(\text{NIP})} \\ \mathbf{m}^{(v)} \end{pmatrix} , \tag{4.15}$$

the cost function with the additional regularization term reads

$$S(\mathbf{m}) = \frac{1}{2}\Delta\mathbf{d}^T(\mathbf{m})\underline{\mathbf{C}}_D^{-1}\Delta\mathbf{d}(\mathbf{m}) + \frac{1}{2}\varepsilon''\mathbf{m}^{(v)T}\underline{\mathbf{D}}''\mathbf{m}^{(v)} . \tag{4.16}$$

The matrix $\underline{\mathbf{D}}''$ is derived in Appendix C. It is positive definite, therefore the regularization term can be interpreted as a squared norm of the velocity model vector $\mathbf{m}^{(v)}$, multiplied by a factor $\varepsilon''$. This weighting factor balances the relative contributions of the data misfit term and the regularization term to the cost function. Note that the minimum curvature constraint is applied to the entire velocity model represented by the vector $\mathbf{m}^{(v)} = \mathbf{m}_n^{(v)} + \Delta\mathbf{m}^{(v)}$, not only to the model update $\Delta\mathbf{m}^{(v)}$.

Again taking the gradient of the cost function $S$ with respect to the model parameters and equating it to zero yields

$$\left(\mathbf{F}^T\underline{\mathbf{C}}_D^{-1}\mathbf{F} + \varepsilon''\underline{\tilde{\mathbf{D}}}''\right)\Delta\mathbf{m} = \mathbf{F}^T\underline{\mathbf{C}}_D^{-1}\Delta\mathbf{d}(\mathbf{m}_n) - \varepsilon''\underline{\tilde{\mathbf{D}}}''\mathbf{m}_n , \tag{4.17}$$

which needs to be solved to obtain the model update vector $\Delta\mathbf{m}$. In equation (4.17), a symmetrical matrix $\underline{\tilde{\mathbf{D}}}''$ has been introduced which is obtained from matrix $\underline{\mathbf{D}}''$ by adding extra rows and columns containing only zeros, such that the regularization term may be written in terms of the complete model vector $\mathbf{m}$:

$$\mathbf{m}^T\underline{\tilde{\mathbf{D}}}''\mathbf{m} = \mathbf{m}^{(v)T}\underline{\mathbf{D}}''\mathbf{m}^{(v)} . \tag{4.18}$$

Equation (4.17) is equivalent to the Gauss-Newton update formulas given by Tarantola (1987). If a matrix $\hat{\mathbf{F}}$ and a vector $\Delta\hat{\mathbf{d}}$ with

$$\hat{\mathbf{F}} = \begin{pmatrix} \underline{\mathbf{C}}_D^{-1/2}\mathbf{F} \\ \underline{\tilde{\mathbf{B}}} \end{pmatrix} \quad \text{and} \quad \Delta\hat{\mathbf{d}} = \begin{pmatrix} \underline{\mathbf{C}}_D^{-1/2}\Delta\mathbf{d}(\mathbf{m}_n) \\ -\underline{\tilde{\mathbf{B}}}\mathbf{m}_n \end{pmatrix} \tag{4.19}$$

are defined, equation (4.17) may again be written in the form of the normal equations:

$$\hat{\mathbf{F}}^T \hat{\mathbf{F}} \Delta \mathbf{m} = \hat{\mathbf{F}}^T \Delta \hat{\mathbf{d}} . \tag{4.20}$$

The matrix $\tilde{\mathbf{B}}$ is defined such that $\tilde{\mathbf{B}}^T \tilde{\mathbf{B}} = \varepsilon'' \tilde{\mathbf{D}}''$. From equation (4.20) it follows that $\Delta \mathbf{m}$ can be computed as the least-squares solution to a matrix equation

$$\hat{\mathbf{F}} \Delta \mathbf{m} = \Delta \hat{\mathbf{d}} . \tag{4.21}$$

Note that those rows of matrix $\tilde{\mathbf{B}}$ in $\hat{\mathbf{F}}$ and $\Delta \hat{\mathbf{d}}$ that correspond to the NIP model parameters are zero. For the solution of equation (4.21) matrix $\tilde{\mathbf{B}}$ in equation (4.19) can therefore be replaced by a rectangular matrix $[\mathbf{0}, \mathbf{B}]$, where $\mathbf{B}^T \mathbf{B} = \varepsilon'' \mathbf{D}''$, see Appendix C.

Solving equation (4.21) by making use of the normal equations would imply computing matrix $\hat{\mathbf{F}}^T \hat{\mathbf{F}}$ and vector $\hat{\mathbf{F}}^T \Delta \hat{\mathbf{d}}$ and solving equation (4.20). However, as already indicated, more efficient and numerically stable methods for obtaining the least-squares solution of equation (4.21) exist. Two such methods are briefly discussed in Chapter 5. Once a model update vector $\Delta \mathbf{m}$ has been obtained, the updated model vector can be computed by

$$\mathbf{m}_{n+1} = \mathbf{m}_n + \lambda \Delta \mathbf{m} . \tag{4.22}$$

The factor $\lambda \leq 1$ is necessary to account for the fact that a linearization has been applied in (4.12) and the cost function is, in fact, not a quadratic function of $\mathbf{m}$.

In order to set up the linear system (4.21), to be solved to obtain a model update, the elements of the matrix $\mathbf{F}$, the Fréchet derivatives $F_{ij} = \partial f_i / \partial m_j$, need to be available. These can be calculated during the forward modeling step by applying ray perturbation theory (Section 2.4) along the considered normal rays in the current model. The Fréchet derivative expressions for the 2D case, calculated in ray-centered coordinates are given in Appendix D. The derivation of the corresponding quantities for the 3D case, calculated in Cartesian coordinates, is outlined in Appendix E.

## 4.5 Additional constraints

In addition to the constraint of minimum second derivatives of the velocity distribution, other constraints on the velocity model may be introduced to further reduce the ambiguity of the inversion problem.

### 4.5.1 A priori velocity information

If the true subsurface velocity is known or can be estimated at certain locations, this information can be used in the tomographic inversion. Such information may either stem from direct measurements obtained, for instance, in boreholes or small-scale near-surface seismic surveys, or from a priori geological knowledge. In the case of marine seismic data, an obvious model constraint is that the near-surface velocity must equal the acoustic velocity in water. Such a priori velocity information is treated in the context of the tomographic inversion as additional data. If velocity values are available at a total of $n_{\text{vdata}}$ subsurface locations, an additional $n_{\text{vdata}}$ elements

$$v(x_i, y_i, z_i) \qquad i = 1, \ldots, n_{\text{vdata}} \tag{4.23}$$

are added to the data vector $\mathbf{d}$. During the inversion, the misfit between these velocity values and the model velocities at the respective subsurface points is minimized. The corresponding diagonal elements of the matrix $\underline{\mathbf{C}}_D$ determine the relative weight given to these additional data points. Note that even if the velocity information contained in these data points is considered reliable, a certain residual data error must be allowed, as this information may be incompatible with the smooth model description used in the inversion. The additional elements of the tomographic matrix $\underline{\mathbf{F}}$ associated with the data points (4.23), that is, the corresponding Fréchet derivatives, are derived in Appendix F. Obviously, these derivatives involve only the model parameters associated with the velocity model itself (the B-spline coefficients).

### 4.5.2 Velocity and reflector structure

Another type of additional constraint on the model comes from the assumption that the velocity structure should locally follow the reflector structure. In other words, velocity variations should mainly occur in the direction perpendicular to reflectors. Such an assumption is often reasonable, as velocity changes are likely to occur mainly between different geological layers rather than along the layers. The true subsurface reflector structure is initially unknown, as it depends on the velocity structure to be determined. However, in a given velocity model, the orientations of reflector elements are known at all NIP locations corresponding to the data points (4.6) used in the inversion. The direction in which velocity variations should mainly occur, locally coincides with the normal ray direction at the respective NIP. Conversely, velocity variations in the plane locally perpendicular to that direction should be small. The velocity structure can, thus, be forced to locally follow the reflector structure by minimizing the norm of the velocity gradient in the plane perpendicular to the normal ray at each considered NIP (Figure 4.4). If the upwards-pointing reflector normal at the NIP is defined in the general 3D case by a unit vector $\hat{\mathbf{e}}^{(\text{NIP})} = (e_x^{(\text{NIP})}, e_y^{(\text{NIP})}, e_z^{(\text{NIP})})^T$ (compare Section 4.3), two additional vectors $\hat{\mathbf{e}}_1$ and $\hat{\mathbf{e}}_2$ can easily be constructed such that $\hat{\mathbf{e}}_1$ and $\hat{\mathbf{e}}^{(\text{NIP})}$ define a vertical plane through the NIP and $\hat{\mathbf{e}}_1$, $\hat{\mathbf{e}}_2$, and $\hat{\mathbf{e}}^{(\text{NIP})}$ are mutually perpendicular. The velocity gradient in the local reflector plane is, thus, given by

$$\nabla_{\mathbf{q}} v = \left( \hat{\mathbf{e}}_1 \cdot \nabla v, \; \hat{\mathbf{e}}_2 \cdot \nabla v \right)^T . \tag{4.24}$$

The constraint that the velocity structure should locally follow the reflector structure can, therefore, be imposed by minimizing

$$|\nabla_{\mathbf{q}} v|_i = \left[ (\hat{\mathbf{e}}_1 \cdot \nabla v)^2 + (\hat{\mathbf{e}}_2 \cdot \nabla v)^2 \right]_i^{\frac{1}{2}} \qquad i = 1, \ldots, n_{\text{data}} \tag{4.25}$$

at all considered NIP locations in the current model. This provides another $n_{\text{data}}$ constraints on the model parameters, resulting in as many additional rows in the tomographic matrix. The corresponding Fréchet derivatives for the 2D and 3D case are derived in Appendix F. However, for

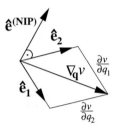

Figure 4.4: In order to force the velocity structure to locally follow the reflector structure, the norm of the gradient of velocity $\nabla_{\mathbf{q}} v$ in the local reflector plane at each considered NIP is minimized.

simplicity, only the Fréchet derivatives with respect to the velocity model parameters are considered. This means that the influence of variations of the NIP model parameters (4.2) on the computed values of (4.25) is neglected.

The criterion of minimum first derivatives of velocity along the reflector at the NIPs can obviously only be used if the picked data points correspond to reflection events in the data. For diffraction events this criterion makes no sense. Even for reflection events the criterion should be used with care as it is based on a somewhat crude assumption. Therefore, a certain residual error should be allowed in minimizing (4.25) and the corresponding weights (the diagonal elements of matrix $\underline{\mathbf{C}}_D$) should be chosen accordingly.

### 4.5.3  Spatially variable model smoothness

Another way of introducing a priori information on the velocity distribution involves the regularization term (4.18) (the second term in the cost function (4.16)) itself. As discussed in Section 4.4 and in detail in Appendix C, this term represents a measure of the overall curvature of the velocity model in terms of its second spatial derivatives. If it is a priori known that in certain parts of the model, velocities should be especially smooth, or if parts of the model are not constrained by the data (4.6) at all, it is useful to apply a spatially variable regularization. This is achieved by using spatially variable weights in the calculation of the model curvature measure. For the general 3D case it then reads

$$\int_z \int_y \int_x \left[ \varepsilon_{xx}(x,y,z) \left( \frac{\partial^2 v(x,y,z)}{\partial x^2} \right)^2 + \varepsilon_{yy}(x,y,z) \left( \frac{\partial^2 v(x,y,z)}{\partial y^2} \right)^2 \right.$$
$$\left. + \varepsilon_{zz}(x,y,z) \left( \frac{\partial^2 v(x,y,z)}{\partial z^2} \right)^2 + \varepsilon\, v^2(x,y,z) \right] dx\,dy\,dz . \qquad (4.26)$$

Practical experience has shown that this spatially variable smoothness constraint is especially useful at the borders of the velocity model, where there may not be sufficient information in the data (4.6) to sufficiently constrain the outermost B-spline coefficients. The implementation of the spatially varying weights is discussed in Appendix C.

# 4.6 Inversion algorithm

In the previous sections, the essential elements of tomographic inversion with kinematic wavefield attributes have been presented. The step-by-step procedure described in the following combines these elements into a general inversion scheme which is valid for 1D, 2D, and 3D inversion. Implementation aspects for these different cases and details of the algorithm will be discussed in Sections 5.1, 5.2, and 5.3. Here, it is assumed, that the appropriate data components for the considered case have been extracted from the seismic data and are available as input for the inversion. The algorithm proceeds as follows:

1) An initial velocity model is set up by assigning meaningful values to the elements of $\mathbf{m}^{(v)}$. Data weights (the diagonal elements of matrix $\underline{\mathbf{C}}_D$), additional constraints (Section 4.5) and regularization weights also need to be specified.

2) For each data point, a normal ray is traced into the model until the respective traveltime $\tau_0$ is used up. Each ray is started at the measurement surface ($z = 0$) at a location and with an initial ray direction defined by the corresponding data components. The ray end point in the subsurface, defined by $\tau_0 = 0$, is the initial NIP location. The ray slowness vector at the NIP yields the local reflector normal. With this information, the initial NIP model vector $\mathbf{m}^{(\mathrm{NIP})}$ can be set up.

3) Forward modeling is performed by dynamic ray tracing in the upward direction, starting at the respective NIP, to obtain the elements of $\mathbf{d}_{\mathrm{mod}}$. Simultaneously, the elements of the matrix $\underline{\mathbf{F}}$ (the Fréchet derivatives) are computed by applying ray perturbation theory along each normal ray.

4) The cost function (4.16) is evaluated by calculating the data misfit from $\mathbf{d}$ and $\mathbf{d}_{\mathrm{mod}}$, and calculating the regularization term from the velocity model vector $\mathbf{m}^{(v)}$.

5) The linear system of equations (4.21) is set up and solved in the least-squares sense with an appropriate numerical method to obtain a model update vector $\Delta\mathbf{m}$.

6) The model update multiplied by a factor $\lambda \leq 1$ is added to the current model and forward modeling (dynamic ray tracing) is performed with the new model parameters to obtain a new vector $\mathbf{d}_{\mathrm{mod}}$.

7) The cost function with the data misfit calculated from $\mathbf{d}$ and the new $\mathbf{d}_{\mathrm{mod}}$ is evaluated.

8) If the cost function has increased, the updated model is rejected, $\lambda$ is decreased and steps 6) and 7) are repeated until the cost function decreases or $\lambda$ falls below a specified value. If the cost function does not decrease, even for small values of $\lambda$, a minimum of the cost function has been reached.

9) If the cost function has decreased, the updated model is accepted, the regularization weight $\varepsilon''$ is decreased and the next iteration is started by going back to step 3) with the new model. The procedure is stopped if a specified maximum number of iterations has been reached or if the cost function has fallen below a certain specified value

Except for step 1), all described steps of the inversion algorithm are performed automatically with no human interaction.

The systematic decrease of the regularization weight $\varepsilon''$ during the inversion has the effect of allowing the long-wavelength features of the velocity model to be determined during the early stages of the inversion process, while more and more model details can be resolved in later iterations. As discussed by Williamson (1990) and Nemeth et al. (1997), the application of such a dynamic regularization scheme during tomographic inversion leads to an improved solution stability and significantly accelerates the convergence of the process. Different strategies for decreasing the regularization weight in the course of iterations can be applied. In the tomography implementations discussed in Chapter 5, a simple rule for the decrease of $\varepsilon''$ based on the decrease of the overall value of the cost function will be used.

# Chapter 5

# Implementation

In the previous chapter, the general concept of velocity model estimation with kinematic wave-field attributes related to emerging NIP waves has been introduced. Based on this concept, a tomographic inversion algorithm for the determination of smooth velocity models has been formulated for the general 3D case. This tomographic inversion will now be discussed in more detail for the special cases of 1D inversion (Section 5.1), 2D inversion (Section 5.2), and 3D inversion with limited azimuth information (Section 5.3). Starting out with the results of Chapter 4, the general formulation of the tomographic inversion will be adapted to the respective special cases. Different implementational aspects required for the numerical realization of the algorithm will be discussed, and the resulting implementations for all three special cases will be demonstrated and evaluated on synthetic test examples. In Chapter 6, the 2D version of the tomographic inversion will then be applied to synthetic and real seismic data.

## 5.1   1D tomographic inversion

If the subsurface velocity distribution is laterally invariant and reflectors are horizontal, the problem of determining subsurface velocities simplifies significantly. In this situation, known as the 1D case, it is usually sufficient to apply Dix inversion (Dix, 1955). The 1D version of the tomographic inversion introduced in Chapter 4 will be discussed here solely because it allows to demonstrate some of the basic features of the method.

### 5.1.1   Data and model components

In the 1D case, the slowness vector of a normal ray has no horizontal component and the measured traveltimes are independent of the normal ray emergence location. Therefore, two data components are sufficient to characterize an emerging NIP wave and the input for the 1D tomographic inversion consists of $n_{\text{data}}$ data points

$$(\tau_0, M_{\text{NIP}})_i \qquad i = 1, \ldots, n_{\text{data}} , \tag{5.1}$$

which can be extracted from a single CMP gather using the traveltime equation (3.4) in the form

$$t^2 = t_0^2 + 2t_0 M_{\text{NIP}} h^2 \ . \tag{5.2}$$

A NIP in the subsurface associated with a data point is completely determined by its depth, or its vertical coordinate $z^{(\text{NIP})} < 0$. The 1D velocity function $v(z)$ is defined by $n_z$ B-spline coefficients:

$$v(z) = \sum_{k=1}^{n_z} v_k \beta_k(-z) \ . \tag{5.3}$$

Here, the $\beta_k(-z)$ are B-splines of degree $m = 3$ (cubic B-splines). The minus sign accounts for the fact that the positive $z$-direction points upwards, while the B-spline knot sequence is defined to be increasing with depth, see Section 4.3 and Appendix B. The model parameters to be determined during the tomographic inversion are, thus, given by

$$\begin{aligned} z_i^{(\text{NIP})} && i = 1, \ldots, n_{\text{data}} \\ v_k && k = 1, \ldots, n_z \ . \end{aligned} \tag{5.4}$$

Consequently, there are a total of $2n_{\text{data}}$ data components versus $n_{\text{data}} n_z$ model components.

## 5.1.2   Modeling and Fréchet derivatives

The expressions required for performing 1D forward modeling to obtain the quantities $\tau_0$ and $M_{\text{NIP}}$ associated with a NIP in a given model are simple. While the traveltime along the normal ray is obtained by integration of $v^{-1}(z)$ along the ray, the second traveltime derivative of the emerging NIP wave, $M_{\text{NIP}}$, can be computed by dynamic ray tracing using the system of differential equations (2.72) with initial conditions $(0,1)^T$ at $z = z^{(\text{NIP})}$ to obtain $Q_2$ and $P_2$. However, due to $\partial^2 v / \partial x^2 \equiv 0$, the calculation of $Q_2$ and $P_2$ in the 1D case amounts to a simple integration over velocity along the ray to obtain $Q_2$, while $P_2$ remains constant. Consequently, $\tau_0$ and $M_{\text{NIP}}$ can be calculated by numerical integration of

$$\tau_0 = \int_{z^{(\text{NIP})}}^0 v^{-1}(z)\, dz \quad \text{and} \quad M_{\text{NIP}} = \left[ \int_{z^{(\text{NIP})}}^0 v(z)\, dz \right]^{-1} \ . \tag{5.5}$$

The expressions for the Fréchet derivatives in the 1D case are derived in Appendix D as a special case of the corresponding 2D Fréchet derivatives. They read

$$\frac{\partial \tau_{0i}}{\partial z_j^{(\text{NIP})}} = -\frac{1}{v(z_j^{(\text{NIP})})} \delta_{ij} \tag{5.6}$$

$$\frac{\partial \tau_{0i}}{\partial v_k} = -\int_{z_i^{(\text{NIP})}}^0 \frac{\beta_k(-z)}{v^2(z)}\, dz \tag{5.7}$$

$$\frac{\partial M_{\text{NIP}i}}{\partial z_j^{(\text{NIP})}} = v(z_j^{(\text{NIP})}) \left[ \int_{z_j^{(\text{NIP})}}^0 v(z)\, dz \right]^{-2} \delta_{ij} \tag{5.8}$$

$$\frac{\partial M_{\text{NIP}i}}{\partial v_k} = -\int_{z_i^{(\text{NIP})}}^0 \beta_k(-z)\, dz \left[ \int_{z_i^{(\text{NIP})}}^0 v(z)\, dz \right]^{-2} \ . \tag{5.9}$$

Note that the $z$-axis is defined positive upwards. The Kronecker symbol $\delta_{ij}$ in equations (5.6) and (5.8) accounts for the fact that quantities that correspond to different normal rays are independent of each other. The $v_k$ in equations (5.7) and (5.9) is the $k$th B-spline coefficient, associated with the B-spline basis function $\beta_k(-z)$. The derivative with respect to these coefficients is required rather than that with respect to the velocity itself, as the $v_k$ are the model parameters to be determined. Expressions (5.6) to (5.9) can be computed by simple numerical integration in the given model.

### 5.1.3 The tomographic matrix

With the expressions (5.6) to (5.9), the tomographic matrix can be set up. If the data components (5.1) are arranged into a column vector $\mathbf{d}$,

$$\mathbf{d} = (\tau_{01}, \ldots, \tau_{0n_{\text{data}}}, M_{\text{NIP}1}, \ldots, M_{\text{NIP}n_{\text{data}}})^T , \tag{5.10}$$

and the model components are arranged into a model vector $\mathbf{m}$,

$$\mathbf{m} = (z_1^{(\text{NIP})}, \ldots, z_{n_{\text{data}}}^{(\text{NIP})}, v_1, \ldots, v_{n_z})^T , \tag{5.11}$$

matrix $\hat{\mathbf{F}}$ in equation (4.21) takes the form

$$\hat{\mathbf{F}} = \begin{pmatrix} \dfrac{1}{\sigma_\tau}\left[\dfrac{\partial \tau_0}{\partial z^{(\text{NIP})}}\right] & \dfrac{1}{\sigma_\tau}\left[\dfrac{\partial \tau_0}{\partial v}\right] \\[2mm] \dfrac{1}{\sigma_M}\left[\dfrac{\partial M_{\text{NIP}}}{\partial z^{(\text{NIP})}}\right] & \dfrac{1}{\sigma_M}\left[\dfrac{\partial M_{\text{NIP}}}{\partial v}\right] \\[2mm] \hline \left[\,0\,\right] & \left[\,B\,\right] \end{pmatrix} , \tag{5.12}$$

where $\left[\frac{\partial \tau_0}{\partial z^{(\text{NIP})}}\right]$ and $\left[\frac{\partial M_{\text{NIP}}}{\partial z^{(\text{NIP})}}\right]$ represent $n_{\text{data}} \times n_{\text{data}}$ diagonal matrices containing the Fréchet derivatives (5.6) and (5.8), respectively. As noted above, off-diagonal elements in these matrices are zero, as they connect quantities associated with different NIP waves. The matrices $\left[\frac{\partial \tau_0}{\partial v}\right]$ and $\left[\frac{\partial M_{\text{NIP}}}{\partial v}\right]$ are $n_{\text{data}} \times n_z$ matrices containing the Fréchet derivatives (5.7) and (5.9), while matrix $\left[\,B\,\right]$ is an $n_z \times n_z$ upper-triangular matrix identical to matrix $\mathbf{B}$ defined in Appendix C. It is calculated from matrix $\varepsilon'' \mathbf{D}''^{(1\text{D})}$ (Appendix C) by Cholesky decomposition (e. g., Press et al., 1992). Finally, matrix $\left[\,0\,\right]$ represents an $n_z \times n_{\text{data}}$ zero matrix. If no additional constraints on the model (Section 4.5) are applied, the matrix $\hat{\mathbf{F}}$, thus, has dimensions $(2n_{\text{data}} + n_z) \times (n_{\text{data}} + n_z)$. The factors $\sigma_\tau$ and $\sigma_M$ are weighting factors for the different data components. As discussed in Section 4.4, they are required for bringing the numerical values of the different types of data to a comparable size, thus stabilizing the inversion process. It turns out that in typical seismic exploration environments with target depth of a few kilometers, if $\tau_0$ is measured in $10^{-3}$ s and $M_{\text{NIP}}$ is measured in $10^{-9}$ s/m$^2$, the factors $\sigma_\tau$ and $\sigma_M$ may be set equal to one.

### 5.1.4 Solution of the linearized inverse problem

During each iteration, the model update vector $\Delta\mathbf{m}$ is obtained as the least-squares solution of the linear system (4.21). Calculating $\Delta\mathbf{m}$ by solving the normal equations (4.20) requires the explicit calculation of $\hat{\mathbf{F}}^T\hat{\mathbf{F}}$ and $\hat{\mathbf{F}}^T\Delta\hat{\mathbf{d}}$, which involves numerical inaccuracies. These can be avoided by directly computing a generalized inverse of matrix $\hat{\mathbf{F}}$ using *Singular Value Decomposition* (SVD) (Lanczos, 1961).

**Singular Value Decomposition**

With Singular Value Decomposition (SVD), an $n \times p$ matrix $\hat{\mathbf{F}}$ with $n \geq p$ can be factored into a product of three matrices:

$$\hat{\mathbf{F}} = \underline{\mathbf{U}}\mathbf{\Lambda}\underline{\mathbf{V}}^T . \tag{5.13}$$

Here, $\underline{\mathbf{U}}$ is an $n \times p$ matrix whose columns contain $p$ of the $n$ orthonormal eigenvectors of the matrix $\hat{\mathbf{F}}\hat{\mathbf{F}}^T$ spanning the data space, namely those associated with the $p$ non-zero eigenvalues. Matrix $\underline{\mathbf{V}}$ is a $p \times p$ matrix whose columns contain the $p$ orthogonal eigenvectors of $\hat{\mathbf{F}}^T\hat{\mathbf{F}}$ that span the model space. Finally, matrix $\mathbf{\Lambda}$ is a $p \times p$ diagonal matrix containing at most $p$ non-zero positive diagonal elements, the singular values $\lambda_i$. Due to the orthogonality of the columns of the matrices $\underline{\mathbf{U}}$ and $\underline{\mathbf{V}}$, respectively, $\underline{\mathbf{U}}^T\underline{\mathbf{U}} = \underline{\mathbf{V}}^T\underline{\mathbf{V}} = \underline{\mathbf{V}}\,\underline{\mathbf{V}}^T = \mathbf{I}_p$. These matrices allow to write the least-squares solution of (4.21) as

$$\Delta\mathbf{m} = \underline{\mathbf{V}}\mathbf{\Lambda}^{-1}\underline{\mathbf{U}}^T\hat{\mathbf{d}} , \tag{5.14}$$

which is equivalent to the solution of the normal equations as $\underline{\mathbf{V}}\mathbf{\Lambda}^{-1}\underline{\mathbf{U}}^T = (\hat{\mathbf{F}}^T\hat{\mathbf{F}})^{-1}\hat{\mathbf{F}}^T$. Thus, the explicit computation of $\hat{\mathbf{F}}^T\hat{\mathbf{F}}$ with the (possibly very large) condition number $(\max(\lambda_i^2)/\min(\lambda_i^2))$ is avoided.

In addition to allowing a stable and efficient least-squares solution of equation (4.21), SVD also provides information on which model parameter combinations can be determined from the data. Equation (4.21) only contains information on model components which lie in the model subspace spanned by the model eigenvectors associated with non-zero singular values. Model components associated with zero or near-zero singular values cannot be reliably determined. These components can be identified by inspecting the singular value spectrum (see Figure 5.5a for an example). It follows from equation (5.14) that small singular values will have a dominant influence on the solution. Therefore, near-zero singular values that are expected to be numerically inaccurate need to be set to zero (SVD truncation). Alternatively, the regularization can be adjusted to ensure that all model parameters are well constrained.

SVD is used here to solve the 1D tomographic inverse problem. However, for larger inverse problems like those encountered in 2D and 3D tomographic inversion (Sections 5.2 and 5.3), SVD becomes inefficient and alternative methods for the solution of equation (4.21) need to be found.

**Model resolution**

In order to examine, how well the different model parameters can be independently resolved by the least-squares solution of equation (4.21), the model resolution matrix (e. g., Menke, 1984) is employed. For nonlinear problems, the model resolution matrix has been derived by Ory and

Pratt (1995). A perturbation $\Delta \mathbf{m}_{\text{true}}$ of the true solution $\mathbf{m}_{\text{true}}$ of the inverse problem results in a first-order data perturbation of

$$\Delta \mathbf{d}_{\text{obs}} = \underline{\mathbf{F}} \Delta \mathbf{m}_{\text{true}} \ . \tag{5.15}$$

To investigate, how well the true perturbed model $\mathbf{m}_{\text{true}} + \Delta \mathbf{m}_{\text{true}}$ can be reconstructed from this data perturbation, equation (5.15) is inserted into the expression for the estimated model,

$$
\begin{aligned}
\Delta \mathbf{m} &= (\hat{\underline{\mathbf{F}}}^T \hat{\underline{\mathbf{F}}})^{-1} \hat{\underline{\mathbf{F}}}^T \left( \begin{array}{c} \underline{\mathbf{C}}_D^{-1/2} \underline{\mathbf{F}} \Delta \mathbf{m}_{\text{true}} \\ -\tilde{\underline{\mathbf{B}}} \, \mathbf{m}_{\text{true}} \end{array} \right) \\
&= (\hat{\underline{\mathbf{F}}}^T \hat{\underline{\mathbf{F}}})^{-1} (\underline{\mathbf{F}}^T \underline{\mathbf{C}}_D^{-1} \underline{\mathbf{F}} \Delta \mathbf{m}_{\text{true}} - \tilde{\underline{\mathbf{B}}}^T \tilde{\underline{\mathbf{B}}} \, \mathbf{m}_{\text{true}}) \\
&= (\hat{\underline{\mathbf{F}}}^T \hat{\underline{\mathbf{F}}})^{-1} (\underline{\mathbf{F}}^T \underline{\mathbf{C}}_D^{-1} \underline{\mathbf{F}} (\mathbf{m}_{\text{true}} + \Delta \mathbf{m}_{\text{true}}) - \hat{\underline{\mathbf{F}}}^T \hat{\underline{\mathbf{F}}} \mathbf{m}_{\text{true}}) \ .
\end{aligned}
\tag{5.16}
$$

It follows that

$$(\mathbf{m}_{\text{true}} + \Delta \mathbf{m}) = (\hat{\underline{\mathbf{F}}}^T \hat{\underline{\mathbf{F}}})^{-1} \underline{\mathbf{F}}^T \underline{\mathbf{C}}_D^{-1} \underline{\mathbf{F}} (\mathbf{m}_{\text{true}} + \Delta \mathbf{m}_{\text{true}}) \ , \tag{5.17}$$

thus, the resolution matrix is given by

$$\underline{\mathbf{R}} = (\hat{\underline{\mathbf{F}}}^T \hat{\underline{\mathbf{F}}})^{-1} \underline{\mathbf{F}}^T \underline{\mathbf{C}}_D^{-1} \underline{\mathbf{F}} \ . \tag{5.18}$$

It can be calculated if the matrix $\hat{\underline{\mathbf{F}}}^T \hat{\underline{\mathbf{F}}}$ is regular. Unlike the model resolution matrix for linear problems, matrix $\underline{\mathbf{R}}$ in equation (5.18) is evaluated at the true model which is assumed to be known. Alternatively, it can be evaluated at the optimum model determined by nonlinear inversion.

If $\underline{\mathbf{R}}$ is equal to the identity matrix, all model parameters are perfectly resolved. Otherwise, the estimated model parameters represent weighted averages of the true model parameters (Menke, 1984). Note that any deviation of $\underline{\mathbf{R}}$ from an identity matrix is due to the regularization term, as from the definition of $\hat{\underline{\mathbf{F}}}$, equation (4.19),

$$\hat{\underline{\mathbf{F}}}^T \hat{\underline{\mathbf{F}}} = \underline{\mathbf{F}}^T \underline{\mathbf{C}}_D^{-1} \underline{\mathbf{F}} + \varepsilon'' \tilde{\underline{\mathbf{D}}}'' \ . \tag{5.19}$$

Thus, $\underline{\mathbf{R}} = \underline{\mathbf{I}}$ only if $\underline{\mathbf{F}}^T \underline{\mathbf{C}}_D^{-1} \underline{\mathbf{F}}$ is regular and $\varepsilon''$ can be set to zero. The resolution with which the velocity model $v(z)$ itself can be determined depends, however, on the B-spline knot spacing used to define the model. The B-spline basis functions themselves already have a spatial extension (Figure B.1). Reducing the knot spacing decreases the spatial extension of the basis functions, but increases the number of model parameters. This may cause the inverse problem to be underdetermined, making a stronger regularization necessary. The model resolution that may be obtained eventually depends on the amount of data that are available and on their reliability and noise level.

## 5.1.5 Synthetic data example

In order to demonstrate the 1D tomographic inversion and evaluate its performance, it is applied to a synthetic dataset. For that purpose, a model consisting of 14 constant-velocity layers separated by plane horizontal interfaces is defined (Figure 5.1a), in which a CMP gather is modeled by ray tracing (Figure 5.1b). For the modeling, a 30 Hz zero-phase Ricker wavelet is used and band-limited noise is added to the resulting seismic traces. The sampling interval is 2 ms. For each

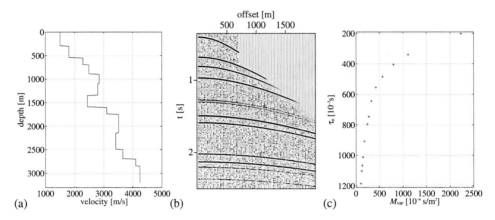

Figure 5.1: 1D synthetic data example. (a) Layered 1D velocity structure. (b) Synthetic CMP gather modeled by ray tracing in the layered 1D model shown in (a). Band-limited noise is added to the data and a mute is applied. (c) Input data for the 1D tomographic inversion: $(M_{NIP}, \tau_0)$ pairs extracted from the CMP gather in (b) with a coherence analysis using equation (5.2). See main text for details.

zero-offset sample, the data components $(\tau_0, M_{NIP})$ are determined by a coherence analysis in the CMP gather using the traveltime equation (5.2). The data components corresponding to reflection events (characterized by coherence maxima) are extracted and are displayed in Figure 5.1c. These 13 data points are used as input for the 1D tomographic inversion with the aim of obtaining a smooth velocity model in which the reflector depths associated with the data points match the true reflector depths (the steps in the velocity distribution in Figure 5.1a). The obtained velocity model should resemble a smoothed version of the true layered model. In this example, the model to be determined in the inversion is described by 15 B-spline coefficients defined at knot locations with a constant vertical spacing of 200 m. Thus, the model vector consists of 28 elements, while the data vector has 26 elements. For the initial model, a near-surface velocity of 1500 m/s and a constant velocity gradient of 2 s$^{-1}$ is used. The normalization and weighting factors in the regularization term are chosen such that $\varepsilon = 0.0001 \varepsilon_{zz}$ in equation (C.2).

The inversion result at different stages of the iterative inversion process (different iteration numbers) is displayed in Figure 5.2. The final smooth model obtained after 12 iterations (bottom right of Figure 5.2) closely resembles the exact model. More importantly, the errors of the reconstructed reflector depths after 12 iterations, displayed in Figure 5.3a, remain below 7 m. This observation again justifies the use of a smooth velocity model description. Obviously, such a model allows to reliably determine reflector depths, even if the true subsurface velocity distribution is not smooth. Figures 5.3b and c show the remaining misfits in the data components $\tau_0$ and $M_{NIP}$, respectively, after 12 iterations. Using the parabolic form of equation (5.2), it is easy to see that an error of $10^{-9}$ s/m$^2$ in $M_{NIP}$ corresponds to a traveltime error of $10^{-3}$ s at a source-receiver offset of 2000 m. The residual data errors are well balanced between $\tau_0$ and $M_{NIP}$.

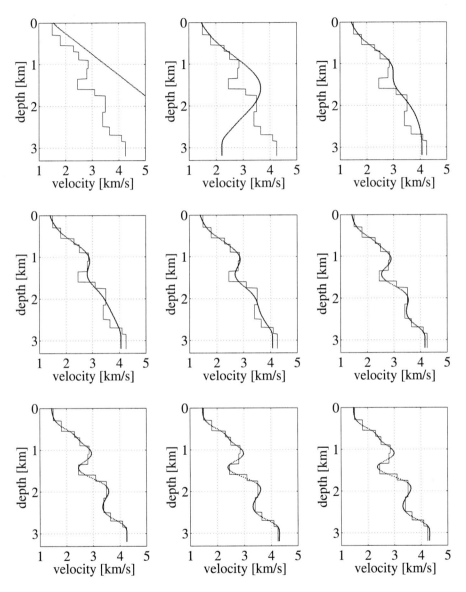

Figure 5.2: 1D synthetic data example. Convergence of the velocity model to the optimum solution. Top left to bottom right: velocity model after 0 (initial model), 1, 2, 3, 4, 6, 8, 10, and 12 iterations. For comparison, the true, layered model is also plotted.

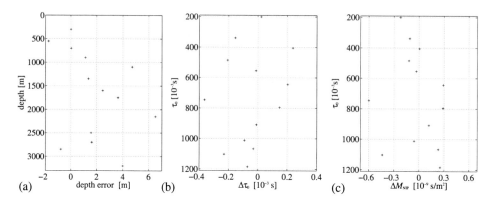

Figure 5.3: 1D synthetic data example. (a) Difference between reflector depths in the true model (steps in the solid line in Figure 5.1a) and the $z^{(\mathrm{NIP})}$ values obtained with the tomographic inversion. (b) Residual data misfit of the data component $\tau_0$ after 12 iterations. (c) Residual data misfit of the data component $M_{\mathrm{NIP}}$ after 12 iterations.

In order to examine the convergence behavior of the inversion process, the value of the cost function $S$ as a function of iteration number is displayed in Figure 5.4. Note that during the inversion, the regularization weight $\varepsilon''$ is gradually decreased from iteration to iteration. As discussed in Section 4.6, such a dynamic regularization accelerates the convergence by first fitting the long-wavelength features of the model and then allowing small-scale details to be resolved in later iterations (Williamson, 1990; Nemeth et al., 1997). This can be well observed in Figure 5.2. Here as well as in the 2D and 3D case, the regularization weight for the $(n+1)$st iteration is given by the heuristic relation

$$\varepsilon''_{n+1} = \sqrt{\frac{S_n}{S_{n-1}}}\, \varepsilon''_n, \qquad (5.20)$$

where $S_n$ is the value of the cost function after $n$ iterations.

In Figures 5.5a and b, the singular value spectrum of $\hat{\mathbf{F}}$ and the corresponding model eigenvectors (the columns of matrix $\underline{\mathbf{V}}$) are displayed for the last iteration step. Obviously, the 13 largest singular values correspond to model eigenvectors that are related mainly to the NIP model parameters (reflector depths). Figure 5.6 shows the resolution matrix $\mathbf{R}$, equation (5.18), for the final model. As expected, the model parameters related to the NIPs, the reflector depths $z^{(\mathrm{NIP})}$, are perfectly resolved as they are not affected by the regularization. The diagonal elements of $\mathbf{R}$ that correspond to the B-spline coefficients have some minor sidelobes, indicating that these parameters could not be perfectly resolved. This is due to the fact that $\varepsilon''$ had to be chosen large enough to ensure the stable computation of the generalized inverse of $\hat{\mathbf{F}}$.

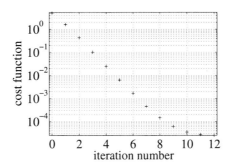

Figure 5.4: 1D synthetic data example. Value of the cost function $S$, equation (4.16), as a function of iteration number. The process converges within 12 iterations.

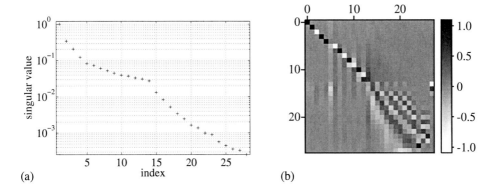

(a)

(b)

Figure 5.5: 1D synthetic data example. (a) Singular value spectrum of the tomographic matrix for the last iteration. The singular values have been sorted according to their magnitude. The 13 largest singular values correspond to model eigenvectors that contain information on reflector depths. (b) Model eigenvectors associated with and sorted according to the singular values in (a). The solution vector $\mathbf{m}$ is a weighted sum of these model eigenvectors.

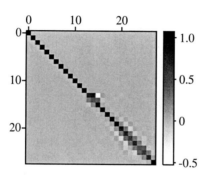

Figure 5.6: 1D synthetic data example. The model resolution matrix, defined in equation (5.18), after the last iteration. The regularization only affects the velocity model parameters. Therefore, the reflector depths are perfectly resolved.

## 5.2   2D tomographic inversion

If subsurface structures and velocities are laterally invariant in one spatial direction (2.5 D case), it is sufficient for the purposes of seismic imaging to record and process seismic data on a seismic line oriented perpendicularly to that direction. Kinematic information for the determination of the velocity model can then be extracted from the seismic data with the 2D CRS stack based on equation (3.7).

### 5.2.1   Data and model components

The parameters describing the second-order traveltimes of emerging NIP wavefronts in the vertical plane defined by the seismic line are the normal ray traveltime $\tau_0 = t_0/2$ and the first and second spatial traveltime derivatives $p^{(\xi)}$ and $M_{\mathrm{NIP}}^{(\xi)}$ at the respective normal ray emergence location $\xi_0$. The data required for the 2D tomographic inversion, thus, consist of data points

$$\left(\tau_0, M_{\mathrm{NIP}}^{(\xi)}, p^{(\xi)}, \xi_0\right)_i \qquad i = 1, \ldots, n_{\mathrm{data}} \tag{5.21}$$

extracted from the results of the 2D CRS stack at $n_{\mathrm{data}}$ pick locations. If the results of the 2D CRS stack are available in the form of emergence angles and wavefront curvatures (or the respective radius values), the required quantities for the data points (5.21) can be obtained from equation (3.8). Each of these data points is associated with a NIP in the subsurface, characterized by its spatial location $(x,z)^{(NIP)}$, where $z^{(NIP)} < 0$, and its local reflector dip $\theta^{(NIP)}$. The 2D velocity model is described by B-splines of degree $m = 4$ (Appendix B):

$$v(x,z) = \sum_{j=1}^{n_x} \sum_{k=1}^{n_z} v_{jk} \beta_j(x) \beta_k(-z) , \tag{5.22}$$

where the $n_x n_z$ coefficients $v_{jk}$ are the velocity model parameters to be determined. For the 2D tomographic inversion, the model is therefore defined by the model parameters

$$\begin{aligned} (x, z, \theta)_i^{(NIP)} \qquad & i = 1, \ldots, n_{\mathrm{data}} \\ v_{jk} \qquad & j = 1, \ldots, n_x , \quad k = 1, \ldots, n_z . \end{aligned} \tag{5.23}$$

There are, thus, a total of $4 n_{\mathrm{data}}$ data components and $(3 n_{\mathrm{data}} + n_x n_z)$ model components.

If additional constraints on the velocity model, as described in Section 4.5, are used in the inversion together with the data components (5.21), these constraints are treated as extra data. The introduction of a priori velocity information at $n_{\mathrm{vdata}}$ spatial locations in the model provides additional data points

$$v(x_i, z_i) \qquad i = 1, \ldots, n_{\mathrm{vdata}} . \tag{5.24}$$

The constraint that the velocity structure should locally follow the reflector structure at the considered NIP locations (implemented as the minimization of the local first velocity derivative along

73

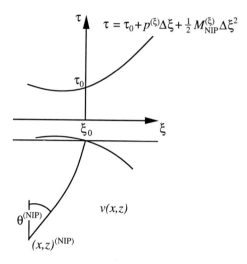

Figure 5.7: Definition of model and data components for 2D tomographic inversion. The data components describe the second-order traveltime curve associated with an emerging NIP wavefront. The corresponding NIP model components are the spatial location of the NIP and the initial normal ray angle, while the velocity model parameters are the B-spline coefficients in equation (5.22).

the reflector) formally provides another $n_{\text{data}}$ data points

$$\left. |\nabla_q v| \right|_{(x,z)_i^{(\text{NIP})}} = 0 \qquad i = 1, \ldots, n_{\text{data}} \, . \tag{5.25}$$

Taking these constraints into account yields a total of $(5 n_{\text{data}} + n_{\text{vdata}})$ data versus $(3 n_{\text{data}} + n_x n_z)$ model components.

### 5.2.2 Modeling and Fréchet derivatives

The forward modeling of the quantities

$$(\tau_0, M_{\text{NIP}}^{(\xi)}, p^{(\xi)}, \xi_0)_i^{\text{mod}} \qquad i = 1, \ldots, n_{\text{data}} \tag{5.26}$$

corresponding to the data points (5.21) during the inversion process is performed by applying 2D kinematic and dynamic ray tracing. Kinematic ray tracing using the 2D version of equations (2.11) yields the emergence location $\xi_0$ and the horizontal slowness component $p^{(\xi)}$ of the normal ray at $\xi_0$, while integration of equation (2.12) along the normal ray yields the traveltime $\tau_0$. As already discussed in Section 4.3 for the general 3D case, the dynamic ray tracing to calculate $M_{\text{NIP}}^{(\xi)}$ may be performed in different coordinate systems. While using dynamic ray tracing in global Cartesian coordinates with a reduced-Hamiltonian formulation (Section 2.6) is numerically more efficient, ray-centered coordinates (Section 2.5) offer more flexibility.

In the Cartesian reduced-Hamiltonian formulation presented in Section 2.6, the $z$-coordinate is used as the independent parameter along the ray. This has the consequence that turning rays with respect to the $z$ direction have to be excluded. Ray-centered coordinates, on the other hand, do not suffer from this restriction, but require spatial derivatives of velocity to be evaluated in arbitrary directions, which decreases numerical efficiency as more B-spline evaluations are necessary. In two dimensions, computation time is not a critical issue. Therefore, ray-centered coordinates will be used here.

In 2D ray-centered coordinates, the second spatial derivative of the NIP wave traveltime on the central ray is given by equation (2.75),

$$M_{\mathrm{NIP}} = P_2/Q_2 \; . \tag{5.27}$$

Transforming this expression to the Cartesian coordinate system of the measurement surface yields (see equation (A.17))

$$M_{\mathrm{NIP}}^{(\xi)} = (\cos^2 \alpha) \, M_{\mathrm{NIP}} = (\cos^2 \alpha) \, P_2/Q_2 \; , \tag{5.28}$$

where a locally constant near-surface velocity has been assumed. The normal ray emergence angle $\alpha$ is obtained from the kinematic ray tracing, using $p^{(\xi)} = \sin \alpha / v(\xi_0)$, where $v(\xi_0)$ is the near-surface velocity at $\xi_0$. For the numerical solution of the kinematic ray-tracing system (the 2D version of equations (2.11)) and the dynamic ray-tracing system (equations (2.72) with normalized point-source and normalized plane-wave initial conditions), a fourth-order Runge-Kutta scheme (e. g., Press et al., 1992) is used.

In practice, the input data components $M_{\mathrm{NIP}}^{(\xi)}$ used during the inversion have been determined from the seismic data with a finite offset aperture and, therefore, do not strictly represent local quantities attributable to the respective normal ray emergence location. Rather, they describe some averaged moveout curve in the considered aperture. In order to account for this fact during the tomographic inversion, it appears useful to apply a corresponding averaging to the model quantities that determine the value of the forward-modeled counterpart of $M_{\mathrm{NIP}}^{(\xi)}$ and its Fréchet derivatives. These quantities are the second and third spatial velocity derivatives in the direction perpendicular to the considered normal ray. Therefore, these derivatives of velocity (and of velocity perturbations) are locally averaged around the normal ray during the application of dynamic ray tracing and ray perturbation theory. The averaging is performed in the direction normal to the ray in an interval of depth-dependent width, which is decreasing linearly from a specified maximum width at the measurement surface to zero at the NIP.

While this averaging procedure is somewhat heuristic, it has proven to be useful in practice. Its effect on the efficiency of the ray tracing is small, as the required average velocity derivatives can be calculated analytically from the values of lower-order derivatives at the endpoints of the averaging interval.

For the tomographic matrix, the Fréchet derivatives of the modeled data components (5.26) with respect to the model parameters (5.23) are needed. These are calculated during ray tracing by applying ray perturbation theory in ray-centered coordinates. The required expressions are derived in Appendix D. Their final form is given in equations (D.14), (D.32), and (D.36). These expressions need to be computed numerically for each considered normal ray. The derivatives with respect to the velocity model parameters need to be calculated separately for each $v_{jk}$ that may affect the considered normal ray, resulting in a large number of quantities to be computed for each ray.

In addition to the Fréchet derivatives associated with the data components (5.21), the derivatives associated with the constraints (5.24) and (5.25) need to be calculated. The forward modeled quantities corresponding to the a priori velocity information (5.24), that is, the velocity model evaluated at the specified locations $(x_i, z_i)$, depend only on the velocity model parameters $v_{jk}$ and not on the NIP model parameters $(x, z, \theta)_i^{(\text{NIP})}$. Therefore, only the Fréchet derivatives with respect to the parameters $v_{jk}$ are non-zero.

The quantities that need to be evaluated for the constraint given in equation (5.25) depend on the velocity model parameters $v_{jk}$ and on the NIP model parameters $(x, z, \theta)_i^{(\text{NIP})}$. It is, however, reasonable to assume that neither the NIP locations, nor the corresponding local reflector dips change drastically enough from iteration to iteration to have any significant effect on the computed values of $\nabla_q v$. The corresponding Fréchet derivatives with respect to the NIP model parameters are therefore neglected. The Fréchet derivatives associated with the constraints (5.24) and (5.25) are derived in Appendix F. The required expressions are given in equations (F.2) and (F.7).

### 5.2.3  The tomographic matrix

Once the Fréchet derivatives are available, the tomographic matrix can be set up. Its form depends on the form of the data and model vectors. For the most general case (with additional model constraints) the data vector is here defined as

$$\mathbf{d} = \begin{pmatrix} \mathbf{d}^{(\text{NIP})} \\ \mathbf{d}^{(\text{constr})} \end{pmatrix} \tag{5.29}$$

with

$$\mathbf{d}^{(\text{NIP})} = (\tau_{01}, \ldots, \tau_{0n_{\text{data}}}, M_{\text{NIP}1}^{(\xi)}, \ldots, M_{\text{NIP}n_{\text{data}}}^{(\xi)}, p_1^{(\xi)}, \ldots, p_{n_{\text{data}}}^{(\xi)}, \xi_{01}, \ldots, \xi_{0n_{\text{data}}})^T \tag{5.30}$$

and

$$\mathbf{d}^{(\text{constr})} = (v(x_1, z_1), \ldots, v(x_{n_{\text{vdata}}}, z_{n_{\text{vdata}}}), 0_1, \ldots, 0_{n_{\text{data}}})^T, \tag{5.31}$$

while the model vector is defined as

$$\mathbf{m} = \begin{pmatrix} \mathbf{m}^{(\text{NIP})} \\ \mathbf{m}^{(v)} \end{pmatrix} \tag{5.32}$$

with

$$\mathbf{m}^{(\text{NIP})} = (x_1^{(\text{NIP})}, \ldots, x_{n_{\text{data}}}^{(\text{NIP})}, z_1^{(\text{NIP})}, \ldots, z_{n_{\text{data}}}^{(\text{NIP})}, \theta_1^{(\text{NIP})}, \ldots, \theta_{n_{\text{data}}}^{(\text{NIP})})^T \tag{5.33}$$

and

$$\mathbf{m}^{(v)} = (v_{11}, \ldots, v_{1n_z}, \ldots, v_{n_x n_z})^T, \tag{5.34}$$

with $m_{[(i-1)n_z + j]}^{(v)} = v_{ij}$, see equation (C.6).

The tomographic matrix $\hat{\underline{\mathbf{F}}}$ in equation (4.21) then takes the form

$$
\hat{\underline{\mathbf{F}}} = \left(
\begin{array}{cccc|c}
\frac{1}{\sigma_\tau}\left[\frac{\partial \tau_0}{\partial x^{(\text{NIP})}}\right] & \frac{1}{\sigma_\tau}\left[\frac{\partial \tau_0}{\partial z^{(\text{NIP})}}\right] & \frac{1}{\sigma_\tau}\left[\frac{\partial \tau_0}{\partial \theta^{(\text{NIP})}}\right] & \frac{1}{\sigma_\tau}\left[\frac{\partial \tau_0}{\partial v}\right] \\
\frac{1}{\sigma_M}\left[\frac{\partial M_{\text{NIP}}^{(\xi)}}{\partial x^{(\text{NIP})}}\right] & \frac{1}{\sigma_M}\left[\frac{\partial M_{\text{NIP}}^{(\xi)}}{\partial z^{(\text{NIP})}}\right] & \frac{1}{\sigma_M}\left[\frac{\partial M_{\text{NIP}}^{(\xi)}}{\partial \theta^{(\text{NIP})}}\right] & \frac{1}{\sigma_M}\left[\frac{\partial M_{\text{NIP}}^{(\xi)}}{\partial v}\right] \\
\frac{1}{\sigma_p}\left[\frac{\partial p^{(\xi)}}{\partial x^{(\text{NIP})}}\right] & \frac{1}{\sigma_p}\left[\frac{\partial p^{(\xi)}}{\partial z^{(\text{NIP})}}\right] & \frac{1}{\sigma_p}\left[\frac{\partial p^{(\xi)}}{\partial \theta^{(\text{NIP})}}\right] & \frac{1}{\sigma_p}\left[\frac{\partial p^{(\xi)}}{\partial v}\right] \\
\frac{1}{\sigma_\xi}\left[\frac{\partial \xi_0}{\partial x^{(\text{NIP})}}\right] & \frac{1}{\sigma_\xi}\left[\frac{\partial \xi_0}{\partial z^{(\text{NIP})}}\right] & \frac{1}{\sigma_\xi}\left[\frac{\partial \xi_0}{\partial \theta^{(\text{NIP})}}\right] & \frac{1}{\sigma_\xi}\left[\frac{\partial \xi_0}{\partial v}\right] \\
\hline
\multicolumn{3}{c|}{\left[\ 0_v\ \right]} & \frac{1}{\sigma_v}\left[\frac{\partial v^{(\text{constr})}}{\partial v}\right] \\
\multicolumn{3}{c|}{\left[\ 0_{v_q}\ \right]} & \frac{1}{\sigma_{v_q}}\left[\frac{\partial (\nabla_q v)}{\partial v}\right] \\
\multicolumn{3}{c|}{\left[\ 0_B\ \right]} & \left[\ B\ \right]
\end{array}
\right) .
\qquad (5.35)
$$

Here, the quantities in the upper left part of $\hat{\underline{\mathbf{F}}}$ are diagonal $n_{\text{data}} \times n_{\text{data}}$ matrices containing the Fréchet derivatives with respect to the NIP model parameters, given in equations (D.14), (D.32), and (D.36). The quantities in the upper right part of $\hat{\underline{\mathbf{F}}}$ are $n_{\text{data}} \times n_x n_z$ matrices containing the corresponding Fréchet derivatives with respect to the velocity model parameters $v_{jk}$.

The $n_{\text{vdata}} \times n_x n_z$ matrix $\left[\frac{\partial v^{(\text{constr})}}{\partial v}\right]$ contains the derivatives associated with the a priori velocity constraints (5.24), given in equation (F.2). The corresponding derivatives with respect to the NIP model parameters are zero, resulting in the $n_{\text{vdata}} \times 3 n_{\text{data}}$ zero matrix $\left[\ 0_v\ \right]$. The elements of the $n_{\text{data}} \times n_x n_z$ matrix $\left[\frac{\partial (\nabla_q v)}{\partial v}\right]$ are the Fréchet derivatives associated with the constraints (5.25) with respect to the velocity model parameters $v_{jk}$. They are given in equation (F.7). As discussed above, the corresponding derivatives with respect to the NIP model parameters are here assumed to be zero, resulting in the $n_{\text{data}} \times 3 n_{\text{data}}$ matrix $\left[\ 0_{v_q}\ \right]$.

Finally, the matrix $\left[\ B\ \right]$ is an $n_x n_z \times n_x n_z$ upper-triangular matrix identical to matrix $\underline{\mathbf{B}}$ defined in equation (C.14). It is calculated from the matrix $\varepsilon'' \underline{\mathbf{D}}''^{(\text{2D})}$ by Cholesky decomposition. The matrix $\left[\ 0_B\ \right]$ is an $n_x n_z \times 3 n_{\text{data}}$ zero matrix.

The weighting or scaling factors $\sigma_\tau$, $\sigma_M$, $\sigma_p$, and $\sigma_\xi$ have been discussed in Section 4.4. They are applied to balance the different involved data types and account for the fact that they have different physical dimensions. If the traveltime $\tau_0$ is measured in $10^{-3}$ s, the second traveltime derivative $M_{\text{NIP}}^{(\xi)}$ is measured in $10^{-9}$ s/m$^2$, the first traveltime derivative $p^{(\xi)}$ is measured in $10^{-6}$ s/m and $\xi_0$ is measured in m, suitable values for the scaling factors are $\sigma_\tau = 1$, $\sigma_M = 1$, $\sigma_p = 2$, and $\sigma_\xi = 1$.

The factors $\sigma_v$ and $\sigma_{v_q}$ are weights for the additional constraints on the velocity model. They determine, how much those constraints contribute to the solution of the inverse problem. The choice of suitable values for these factors depends on how reliable the corresponding a priori model information is. A certain residual error should always be allowed to account for the fact that the a priori velocity values may be incompatible with the smooth velocity model description used in the inversion. Practical experience suggests that if velocities are measured in m/s, a useful value for the corresponding weighting factor is $\sigma_v = 1$.

According to equation (4.19), all of the discussed data weights also need to be applied to the data misfit vector $\Delta \mathbf{d}$ to obtain $\Delta \hat{\mathbf{d}}$ in equation (4.21).

### 5.2.4   Solution of the linearized inverse problem—LSQR algorithm

The linear system of equations obtained from substituting the matrix (5.35) for $\hat{\mathbf{F}}$ in equation (4.21) is in general too large to be solved efficiently by Singular Value Decomposition. On the other hand, $\hat{\mathbf{F}}$ is a sparse matrix, as all of the submatrices in the upper left part of $\hat{\mathbf{F}}$ in equation (5.35) are diagonal and because it contains the zero matrices $\left[ 0_v \right]$, $\left[ 0_{v_q} \right]$, and $\left[ 0_B \right]$.

A suitable method for efficiently solving large sparse linear systems of the form of equation (4.21) in the least-squares sense is the LSQR method introduced by Paige and Saunders (1982a,b). It is an iterative method which is analytically equivalent to the conjugate-gradient algorithm (e. g., Gill et al., 1981), but has more favorable numerical properties, in particular for ill-conditioned systems. The LSQR algorithm allows the matrix $\hat{\mathbf{F}}$ to be stored in a sparse matrix format and directly solves equation (4.21) for $\Delta \mathbf{m}$ in the least-squares sense without explicitly performing any matrix inversions.

The LSQR algorithm itself will here not be described in detail. Different aspects of the method are discussed in the original publications by Paige and Saunders (1982a,b) and in van der Sluis and van der Horst (1987) and Nolet (1987), who also presents a simplified version of the LSQR code. This simplified version of LSQR avoids the need to specify a large number of parameters as required for the application of the algorithm originally published in Paige and Saunders (1982b). It forms the basis of the code used in the tomographic inversion examples in this and the following chapter.

Paige and Saunders (1982a,b) provide several stopping criteria to limit the number of iterations to be performed by the algorithm. The heuristic criterion used here is based on an estimate of the condition number of $\hat{\mathbf{F}}$, provided by the algorithm during each iteration (Paige and Saunders, 1982a, define a condition number also for rectangular matrices). Stopping the iterative procedure as soon as the estimated condition number exceeds some predefined value, effectively acts as a regularization. This approach bears some resemblance to singular value truncation. For the inversion examples presented in this and the following chapter, the limit for the condition number is $10^4$.

### 5.2.5   Synthetic test example

In Chapter 6 the entire velocity model estimation process based on the 2D tomographic inversion with kinematic wavefield attributes is demonstrated on a synthetic and a real seismic dataset.

Here, however, the inversion algorithm is applied to input data that have been directly obtained by dynamic ray tracing along normal rays in a model specified by model parameters of the form of equation (5.23). This allows to demonstrate the performance of the algorithm with perfect input data and to examine the effect of noise in the data on the inversion result.

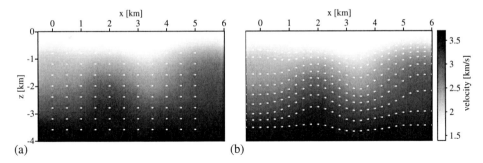

Figure 5.8: 2D synthetic data example. (a) Smooth velocity model defined by B-splines using $11 \times 10$ B-spline coefficients. The knot locations to which the B-spline coefficients are assigned are plotted in white. (b) The NIP locations used to model the input data. They are placed in the model along iso-velocity curves.

**Generating input data**

In order to obtain a test dataset, a laterally inhomogeneous 2D velocity model is defined by specifying B-spline coefficients $v_{jk}$ on a grid of $n_x \times n_z = 11 \times 10$ knot locations with a constant horizontal spacing of 500 m and a constant vertical spacing of 400 m, see Figure 5.8a. In this velocity model, a total of 270 NIP model points are placed in such a way, that the NIP locations and associated reflector dips locally follow the velocity structure, that is, the reflector dips are oriented along iso-velocity curves. For each of these NIP model points, Figure 5.8b, a data point is modeled by dynamic ray tracing along the respective normal ray.

The resulting data components $\tau_0$, $M_{\text{NIP}}^{(\xi)}$, and $p^{(\xi)}$ are displayed in Figures 5.9a, c, and e as a function of the emergence location $\xi_0$ (the fourth data component). Note that the data component $M_{\text{NIP}}^{(\xi)}$ also takes on negative values, indicating the presence of caustics along the respective normal rays. These data points serve as the input for the tomographic inversion with the aim of reconstructing the true smooth velocity distribution. Note that for the application of the tomographic inversion it is in general not necessary for the local reflector dips at the considered NIPs to follow the velocity structure. The reason for placing the NIPs in such a way in this example is that it allows to apply the corresponding additional constraint (Section 4.5) in the inversion and to compare inversion results with and without this constraint.

**Inversion**

For the inversion, a starting velocity model with a near-surface velocity value of 2000 m/s and a vertical velocity gradient of 0.3 s$^{-1}$ is defined. In accordance with the inversion procedure described in Section 4.6, initial NIP model parameters are found automatically by tracing a ray into the subsurface for each data point until the corresponding traveltime $\tau_0$ is used up. The resulting ray endpoint and associated local ray direction provide the initial NIP parameters, that is, the initial NIP location and local reflector dip associated with the considered data point.

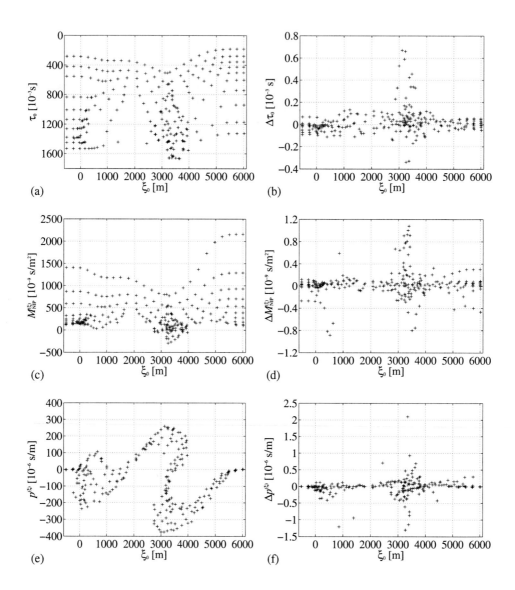

Figure 5.9: 2D synthetic data example. Input data and residual data errors. (a) Values of the data component $\tau_0$ used in the tomographic inversion. (b) Residual errors in $\tau_0$ after 12 iterations. (c) Values of the data component $M_{\mathrm{NIP}}^{(\xi)}$ used in the tomographic inversion. (d) Residual errors in $M_{\mathrm{NIP}}^{(\xi)}$ after 12 iterations. (e) Values of the data component $p^{(\xi)}$ used in the tomographic inversion. (f) Residual errors in $p^{(\xi)}$ after 12 iterations. See main text for a discussion of these results.

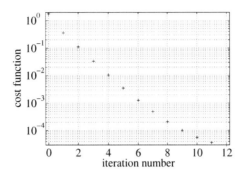

Figure 5.10: 2D synthetic data example. Value of the cost function (4.16) as a function of iteration number. The process converges within 12 iterations.

To better constrain the velocity model at the model boundaries, spatially variable regularization weights as described in Section 4.5 and in Appendix C are used. In this example, the coefficients $\varepsilon_{ij}^{xx}$ in equation (C.15) at the model boundaries ($i = 1$, $i = n_x$, $j = n_z$) are chosen to be a factor 100 larger than those inside the model. Also, the constraint that the velocity structure should follow the reflector structure (minimum velocity gradient along the reflector at each considered NIP location), as discussed above, is used. For the corresponding weighting factor a value of $\sigma_{v_q} = 10$ is chosen (where $\nabla_q v$ is measured in $s^{-1}$).

A total of 12 nonlinear inversion iterations are performed. Each of these iterations involves the least-squares solution of the linear system (4.19) with the LSQR method, which is, in turn, an iterative process.

Following the inversion procedure described in Section 4.6, the regularization weight $\varepsilon''$ is decreased systematically from iteration to iteration to improve the convergence behavior. As in the 1D case, this decrease is controlled by equation (5.20). The value of the cost function $S$, equation (4.16), as a function of iteration number is plotted in Figure 5.10.

**Results**

The residual errors in the data components $\tau_0$, $M_{NIP}^{(\xi)}$, and $p^{(\xi)}$ after 12 iterations are shown in Figures 5.9b, d, and f. In order to assess these residual data errors, they need to be interpreted in terms of the magnitudes of measurement errors that would be expected if the data components were determined from seismic prestack data with a coherence analysis and extracted from the CRS stack results at selected pick locations.

The residual error in $\tau_0$ is clearly less than the expected picking error. Using the parabolic traveltime equation (A.9) it can be shown that an error in $M_{NIP}^{(\xi)}$ of $10^{-9}$ s/m$^2$ corresponds to a traveltime error of $10^{-3}$ s at a source-receiver offset of 2000 m. From the definition of $p^{(\xi)}$ it follows that an error in $p^{(\xi)}$ of $10^{-6}$ s/m corresponds to a traveltime error of $4 \cdot 10^{-4}$ s between the midpoint

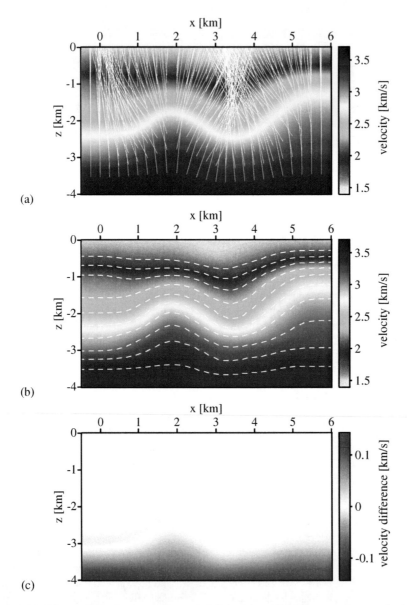

Figure 5.11: 2D synthetic data example. (a) Original laterally inhomogeneous velocity model. The normal rays associated with the true NIP locations (Figure 5.8b) are plotted in white. (b) The inversion result after 12 iterations. Along with the velocity model, the NIP locations and associated reflector dips (represented here as reflector elements) are obtained. (c) Difference section obtained by subtracting the inversion result (b) from the original velocity model. Differences are visible only in the lowermost part of the model, which is not well covered by rays.

aperture endpoints if an aperture of $\Delta\xi = 200$ m is used to determine $p^{(\xi)}$ from the seismic data. Consequently, in this example, the residual errors in all data components could be reduced to a level that lies clearly below the expected measurement error (possible sources of data errors in practice are discussed below).

The reconstructed velocity model itself is displayed in Figure 5.11b together with reconstructed NIP points represented by local reflector elements, or dip bars. For comparison, the true velocity model, together with the normal rays corresponding to the considered NIPs in the model, is displayed in Figure 5.11a. Obviously, the velocity model could be well reconstructed. The difference between the true velocity model and the inversion result is shown in Figure 5.11c. It reveals that differences mainly occur in the lowermost part of the model, which is not constrained by the data, as it is not covered by rays (see Figure 5.11a). To evaluate how well the NIP locations could be reconstructed by the tomographic inversion, Figure 5.12a shows the NIP locations obtained as a result of the tomographic inversion (+) plotted together with the true NIP locations (o). Again, differences are very small and occur mainly in the lowermost part of the model, as shown in Figure 5.12b.

All results discussed so far have been obtained by performing the tomographic inversion with the additional constraint of minimum local velocity gradients along the reflector at all considered NIP locations. It is, of course, also possible to perform the tomographic inversion without this constraint. The resulting velocity model (not displayed) again looks similar to the true model. However, with increasing depth it deviates stronger from the true model than the result obtained with the additional constraint. This also affects the positioning of the reconstructed NIP locations. These are displayed in Figure 5.12c, where they are denoted by (+), together with the true NIP locations (o). The corresponding depth errors are shown in Figure 5.12d.

**Noisy input data**

Up to now, perfect, error-free input data have been used to perform the tomographic inversion. However, in real applications of the method, the input data for the inversion are derived from seismic data. This involves the application of the CRS stack, that is, coherence analyses along traveltime surfaces defined by the second-order approximation (3.7), and the subsequent picking of zero-offset samples to determine where the kinematic wavefield attributes are extracted. Data points thus obtained will contain errors for a number of reasons.

The uncertainty in $\tau_0$ is dominated by picking errors in $t_0 = 2\tau_0$. Among other things, these depend on how well the correct phase to be picked can be identified (frequency content, phase changes, noise level). The error in $p^{(\xi)}$ depends on the noise level in the prestack data and on the size of the midpoint aperture. The midpoint location $\xi_0$ itself is usually known to a high precision in practice. The main source of uncertainty is the data component $M_{\text{NIP}}^{(\xi)}$. It is affected by possible non-hyperbolic moveout, the size of the offset aperture (spread-length bias) and the loss of resolution for deep events (see also the discussion in Section 6.1). For strongly non-hyperbolic moveout, $M_{\text{NIP}}^{(\xi)}$ values determined from the seismic data with equation (3.7) can be completely meaningless. The expected magnitude for errors in $M_{\text{NIP}}^{(\xi)}$ is, therefore, difficult to quantify.

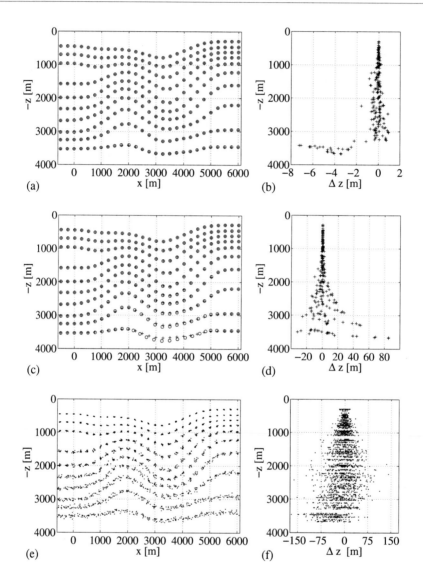

Figure 5.12: 2D synthetic data example. (a) True NIP locations (o) and NIP locations obtained after 12 inversion iterations (+). (b) Depth errors of reconstructed reflection points versus reflector depth. (c) True NIP locations (o) and reconstructed NIP locations without constraint (+). See main text for details. (d) Depth error of reconstructed reflection points without constraint, see (c). (e) Reflection points obtained from tomographic inversions with 10 different noisy datasets, obtained by adding Gaussian noise with standard deviations given in equation (5.36) to the exact data. See main text for details. (f) Depth errors of reconstructed reflection points obtained with noisy datasets.

In any case, it is important to examine the sensitivity and stability of the tomographic inversion with respect to noise in the input data. In the case of linear inverse problems, the effect of errors (with a Gaussian distribution) on the inversion result can be easily evaluated. The model covariance matrix of the final model is linearly related to the data covariance matrix that describes uncertainties in the input data (Menke, 1984).

For nonlinear problems there is no such simple relation. Small data errors may in principle lead to very large errors in the final model. Solutions to the inverse problem may, thus, be unstable. One possible approach for examining the effect of noise in the input data on the inversion result is to generate a large number of noisy datasets by adding different realizations of noise of a given distribution to the exact data and perform the inversion with each of these datasets. In the example presented here, only 10 such noisy datasets are used and it is assumed that data errors in all of the different data components can be described with Gaussian distributions. Although, in view of the different sources of data errors discussed above, the assumption of Gaussian noise is certainly not justified, it is sufficient for evaluating the stability of the tomographic inversion. For that purpose, Gaussian noise with the following standard deviations is added to the different data components:

$$
\begin{aligned}
\sigma_\tau^{\text{noise}} &= 10 \cdot 10^{-3} \text{ s} \\
\sigma_M^{\text{noise}} &= 10 \cdot 10^{-9} \text{ m/s}^2 \\
\sigma_p^{\text{noise}} &= 10 \cdot 10^{-6} \text{ m/s} \\
\sigma_\xi^{\text{noise}} &= 10 \text{ m}
\end{aligned}
\tag{5.36}
$$

An interpretation of errors in $M_{\text{NIP}}^{(\xi)}$ and $p^{(\xi)}$ in terms of traveltime errors at a specified offset or in a specified midpoint aperture has been given above. A total of 10 realizations of Gaussian noise with the standard deviations given in equations (5.36) are generated and added to the data displayed in Figures 5.9a, c, and e. The resulting 10 datasets are used separately to perform the tomographic inversion. All other parameters (number of iterations, regularization weights, scaling factors, additional constraints) are left unaltered.

The NIP locations obtained from the inversions performed with the 10 different datasets are plotted together in Figure 5.12e. The depth errors of the obtained NIP locations relative to the true values are shown in Figure 5.12f. Although there is quite a large scatter in the NIP locations obtained with the noisy data, the overall reflector structure is well reconstructed. The inversion process is, thus, obviously stable with respect to noise in the input data. The scatter in the reconstructed NIP locations is mainly due to the unrealistic assumption of completely uncorrelated high-magnitude traveltime errors. The obtained velocity models themselves are smooth. Consequently, with these velocity models, smooth time-domain reflection events would be transformed into smooth reflector images in depth, irrespective of the reflection point scatter in Figure 5.12e.

## 5.3 3D tomographic inversion

The tomographic inversion with kinematic wavefield attributes has been formulated in Chapter 4 for the general case of 3D inhomogeneous media, assuming that all kinematic wavefield attributes required to describe the second-order traveltimes of emerging NIP wavefronts are available. In practice, however, the seismic acquisition geometry is often limited to a certain azimuth range. As discussed in Section 3.4, the NIP wave second traveltime derivative can in such cases only be obtained in a certain azimuth direction $\phi$. The 3D tomographic inversion will here be discussed in more detail for the case that only the component $M_\phi^{(\xi)}$ of the matrix $\underline{\mathbf{M}}_{\mathrm{NIP}}^{(\xi)}$, associated with the azimuth $\phi$, is available.

### 5.3.1 Data and model components

In the case of 3D acquisition, the emergence location of a normal ray is characterized by two spatial coordinates $\xi_x$ and $\xi_y$. The associated slowness vector at the measurement surface has two horizontal components denoted by $p_x^{(\xi)}$ and $p_y^{(\xi)}$. These provide the first horizontal traveltime derivatives of an emerging hypothetical NIP wave and can be determined from the 3D seismic data irrespective of the azimuth coverage. The quantity $M_\phi^{(\xi)}$ itself is extracted from the seismic data by using equation (3.11), or the corresponding CMP traveltime formula obtained by setting $\Delta \boldsymbol{\xi} = \mathbf{0}$, and substituting $h\hat{\mathbf{e}}_\phi = h(\cos\phi, \sin\phi)^T$ for $\mathbf{h}$. If $\tau_0 = t_0/2$ again denotes the normal ray traveltime, the data points required for the tomographic inversion are then given by

$$(\tau_0, M_\phi^{(\xi)}, p_x^{(\xi)}, p_y^{(\xi)}, \xi_x, \xi_y)_i \qquad i = 1, \ldots, n_{\mathrm{data}}, \qquad (5.37)$$

where the azimuth $\phi$ may be different for each data point but needs to be specified. The data components $M_\phi^{(\xi)}$, $p_x^{(\xi)}$, and $p_y^{(\xi)}$ are taken from the corresponding kinematic wavefield attribute volumes at the selected $n_{\mathrm{data}}$ pick locations $(t_0, \xi_x, \xi_y)$ in the simulated zero-offset volume.

As in the general 3D case described in Section 4.3, each of the data points (5.37) is associated with a NIP in the subsurface, characterized by its spatial coordinates $(x, y, z)^{(\mathrm{NIP})}$ and its local reflector dip, defined by the two horizontal components $e_x^{(\mathrm{NIP})}$ and $e_y^{(\mathrm{NIP})}$ of the unit vector $\hat{\mathbf{e}}^{(\mathrm{NIP})}$ normal to the reflector at the NIP. The 3D velocity model itself is again described by B-splines of degree $m = 4$. It is given by expression (4.3), where the $n_x n_y n_z$ coefficients $v_{jkl}$ are the velocity model parameters to be determined. The model parameters for the 3D tomographic inversion in the case of limited azimuth coverage are, thus, the same as for the general 3D case (Section 4.3):

$$\begin{aligned} (x, y, z, e_x, e_y)_i^{(\mathrm{NIP})} \qquad & i = 1, \ldots, n_{\mathrm{data}} \\ v_{jkl} \qquad & j = 1, \ldots, n_x, \quad k = 1, \ldots, n_y, \quad l = 1, \ldots, n_z. \end{aligned} \qquad (5.38)$$

Consequently, there are a total of $6n_{\mathrm{data}}$ data components and $(5n_{\mathrm{data}} + n_x n_y n_z)$ model components.

Analogously to the 2D case, the additional constraints described in Section 4.5—if they are applied—are treated as extra data. The introduction of a priori velocity information at $n_{\mathrm{vdata}}$ different spatial locations leads to the additional data points

$$v(x_i, y_i, z_i) \qquad i = 1, \ldots, n_{\mathrm{vdata}}. \qquad (5.39)$$

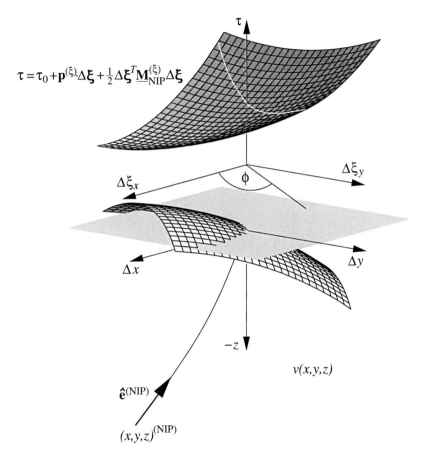

Figure 5.13: In the case of 3D tomographic inversion with limited azimuth coverage, the NIP wave second traveltime derivative is considered in one azimuthal direction $\phi$, only: $M_\phi^{(\xi)} = \hat{\mathbf{e}}_\phi^T \underline{\mathbf{M}}_{\text{NIP}}^{(\xi)} \hat{\mathbf{e}}_\phi$, where $\hat{\mathbf{e}}_\phi = (\cos\phi, \sin\phi)^T$. The NIP model components remain the same as in the general 3D case discussed in Section 4.3. The vector $\Delta\boldsymbol{\xi}$ is defined by $\Delta\boldsymbol{\xi} = \boldsymbol{\xi} - \boldsymbol{\xi}_0$, where $\boldsymbol{\xi}_0$ is the emergence location of the considered normal ray. The horizontal coordinates $\Delta\xi_x$ and $\Delta\xi_y$ are identical to the relative coordinates $\Delta x$ and $\Delta y$.

The constraint of minimum local velocity gradient in the plane of the reflector at each considered NIP location (leading to a velocity structure that locally follows the reflector structure) yields another $n_{data}$ data points

$$\left. |\nabla_q v| \right|_{(x,y,z,e_x,e_y)_i^{(NIP)}} = 0 \qquad i = 1, \ldots, n_{data} \, . \tag{5.40}$$

With these additional constraints the total number of data components becomes $(7 n_{data} + n_{vdata})$.

## 5.3.2 Modeling and Fréchet derivatives

For the forward modeling of the quantities

$$(\tau_0, M_\phi^{(\xi)}, p_x^{(\xi)}, p_y^{(\xi)}, \xi_x, \xi_y)_i^{mod} \qquad i = 1, \ldots, n_{data} \, , \tag{5.41}$$

corresponding to the data points (5.37), the reduced-Hamiltonian formulation in Cartesian coordinates (Section 2.6) with the $z$-coordinate as the running parameter is applied. This has a number of advantages. Firstly, the number of equations for kinematic ray tracing is reduced from 6 as in equations (2.11) to 4 as in equations (2.79). Also, for kinematic as well as for dynamic ray tracing, derivatives of velocity are required in the $x$- and $y$-directions only. Compared to 3D dynamic ray tracing (and ray perturbation theory) in ray-centered coordinates, where velocity derivatives in arbitrary directions may be required, this significantly increases the numerical efficiency, as the number of B-spline evaluations is reduced by a considerable amount. Kinematic ray tracing with equations (2.79) directly yields the required modeled quantities $\xi_x$, $\xi_y$, $p_x^{(\xi)}$, and $p_y^{(\xi)}$ (which are identical to $x_1$, $x_2$, $p_1^{(x)}$, and $p_2^{(x)}$ at the ray endpoint in the notation of Section 2.6), while dynamic ray tracing along the considered normal ray with equation (2.81) allows to directly compute $\underline{\mathbf{M}}_{NIP}^{(\xi)}$ from the elements of the ray propagator matrix (2.82):

$$\underline{\mathbf{M}}_{NIP}^{(\xi)} = \underline{\mathbf{P}}_2^{(x)} \underline{\mathbf{Q}}_2^{(x) \, -1} \, . \tag{5.42}$$

No additional transformations are required, which significantly simplifies the calculation of the associated Fréchet derivatives. The quantity $M_\phi^{(\xi)}$ is obtained from $\underline{\mathbf{M}}_{NIP}^{(\xi)}$ with equation (3.14):

$$M_\phi^{(\xi)} = \hat{\mathbf{e}}_\phi^T \underline{\mathbf{M}}_{NIP}^{(\xi)} \hat{\mathbf{e}}_\phi \, , \tag{5.43}$$

where $\hat{\mathbf{e}}_\phi = (\cos \phi, \sin \phi)^T$.

Using the $z$-coordinate as the running parameter along the ray restricts the applicability of the reduced ray-tracing systems (2.79) and (2.81) to rays that have no turning point with respect to $z$. Turning normal rays rarely occur in practice (in fact, most depth migration algorithms do not handle turning wave energy). If they do occur, they can be easily excluded during the inversion process.

As in the 2D case, a fourth-order Runge-Kutta scheme is used to perform the ray tracing. The Fréchet derivatives of the modeled data components (5.41) with respect to the model parameters (5.38) are again calculated during ray tracing by applying ray perturbation theory. In accordance with the ray tracing itself, ray perturbation theory is also applied using the Cartesian reduced-Hamiltonian formulation of Section 2.6. The required expressions are derived in Appendix E. Again, the Fréchet derivatives with respect to the velocity model parameters need to be calculated for each $v_{jkl}$ that may affect the considered normal ray.

If the additional constraints represented by equations (5.39) and (5.40) are applied, the Fréchet derivatives associated with these constraints also need to be calculated. As in the 2D case, the derivatives associated with the a priori velocity values (5.39) with respect to the NIP model parameters are zero. The dependence of the constraint (5.40) on the NIP model parameters is again neglected so that only the derivatives with respect to the velocity model parameters are considered. The required quantities are derived in Appendix F.

### 5.3.3 The tomographic matrix

With the Fréchet derivatives available, the tomographic matrix can be set up. The data vector is here defined as

$$\mathbf{d} = \begin{pmatrix} \mathbf{d}^{(\mathrm{NIP})} \\ \mathbf{d}^{(\mathrm{constr})} \end{pmatrix} \tag{5.44}$$

with

$$
\begin{aligned}
\mathbf{d}^{(\mathrm{NIP})} = \big( & \tau_{01}, \ldots, \tau_{0n_{\mathrm{data}}}, M^{(\xi)}_{\phi 1}, \ldots, M^{(\xi)}_{\phi n_{\mathrm{data}}}, p^{(\xi)}_{x1}, \ldots, p^{(\xi)}_{x n_{\mathrm{data}}}, \\
& p^{(\xi)}_{y1}, \ldots, p^{(\xi)}_{y n_{\mathrm{data}}}, \xi_{x1}, \ldots, \xi_{x n_{\mathrm{data}}}, \xi_{y1}, \ldots, \xi_{y n_{\mathrm{data}}} \big)^T
\end{aligned}
\tag{5.45}
$$

and

$$\mathbf{d}^{(\mathrm{constr})} = \big( v(x_1, y_1, z_1), \ldots, v(x_{n_{\mathrm{vdata}}}, y_{n_{\mathrm{vdata}}}, z_{n_{\mathrm{vdata}}}), 0_1, \ldots, 0_{n_{\mathrm{data}}} \big)^T , \tag{5.46}$$

while the model vector is defined as

$$\mathbf{m} = \begin{pmatrix} \mathbf{m}^{(\mathrm{NIP})} \\ \mathbf{m}^{(v)} \end{pmatrix} \tag{5.47}$$

with

$$
\begin{aligned}
\mathbf{m}^{(\mathrm{NIP})} = \big( & x^{(\mathrm{NIP})}_1, \ldots, x^{(\mathrm{NIP})}_{n_{\mathrm{data}}}, y^{(\mathrm{NIP})}_1, \ldots, y^{(\mathrm{NIP})}_{n_{\mathrm{data}}}, \\
& z^{(\mathrm{NIP})}_1, \ldots, z^{(\mathrm{NIP})}_{n_{\mathrm{data}}}, e^{(\mathrm{NIP})}_{x1}, \ldots, e^{(\mathrm{NIP})}_{x n_{\mathrm{data}}}, e^{(\mathrm{NIP})}_{y1}, \ldots, e^{(\mathrm{NIP})}_{y n_{\mathrm{data}}} \big)^T
\end{aligned}
\tag{5.48}
$$

and $\mathbf{m}^{(v)}$ defined by

$$m^{(v)}_{[(j-1)n_y n_z + (k-1)n_z + l]} = v_{jkl} \qquad j = 1, \ldots, n_x , \quad k = 1, \ldots, n_y , \quad l = 1, \ldots, n_z . \tag{5.49}$$

The tomographic matrix then takes the form

$$
\hat{\mathbf{F}} =
\begin{pmatrix}
\frac{1}{\sigma_\tau}\left[\frac{\partial \tau_0}{\partial x^{(\mathrm{NIP})}}\right] & \frac{1}{\sigma_\tau}\left[\frac{\partial \tau_0}{\partial y^{(\mathrm{NIP})}}\right] & \frac{1}{\sigma_\tau}\left[\frac{\partial \tau_0}{\partial z^{(\mathrm{NIP})}}\right] & \frac{1}{\sigma_\tau}\left[\frac{\partial \tau_0}{\partial e_x^{(\mathrm{NIP})}}\right] & \frac{1}{\sigma_\tau}\left[\frac{\partial \tau_0}{\partial e_y^{(\mathrm{NIP})}}\right] & \frac{1}{\sigma_\tau}\left[\frac{\partial \tau_0}{\partial v}\right] \\[4pt]
\frac{1}{\sigma_M}\left[\frac{\partial M_\phi^{(\xi)}}{\partial x^{(\mathrm{NIP})}}\right] & \frac{1}{\sigma_M}\left[\frac{\partial M_\phi^{(\xi)}}{\partial y^{(\mathrm{NIP})}}\right] & \frac{1}{\sigma_M}\left[\frac{\partial M_\phi^{(\xi)}}{\partial z^{(\mathrm{NIP})}}\right] & \frac{1}{\sigma_M}\left[\frac{\partial M_\phi^{(\xi)}}{\partial e_x^{(\mathrm{NIP})}}\right] & \frac{1}{\sigma_M}\left[\frac{\partial M_\phi^{(\xi)}}{\partial e_y^{(\mathrm{NIP})}}\right] & \frac{1}{\sigma_M}\left[\frac{\partial M_\phi^{(\xi)}}{\partial v}\right] \\[4pt]
\frac{1}{\sigma_p}\left[\frac{\partial p_x^{(\xi)}}{\partial x^{(\mathrm{NIP})}}\right] & \frac{1}{\sigma_p}\left[\frac{\partial p_x^{(\xi)}}{\partial y^{(\mathrm{NIP})}}\right] & \frac{1}{\sigma_p}\left[\frac{\partial p_x^{(\xi)}}{\partial z^{(\mathrm{NIP})}}\right] & \frac{1}{\sigma_p}\left[\frac{\partial p_x^{(\xi)}}{\partial e_x^{(\mathrm{NIP})}}\right] & \frac{1}{\sigma_p}\left[\frac{\partial p_x^{(\xi)}}{\partial e_y^{(\mathrm{NIP})}}\right] & \frac{1}{\sigma_p}\left[\frac{\partial p_x^{(\xi)}}{\partial v}\right] \\[4pt]
\frac{1}{\sigma_p}\left[\frac{\partial p_y^{(\xi)}}{\partial x^{(\mathrm{NIP})}}\right] & \frac{1}{\sigma_p}\left[\frac{\partial p_y^{(\xi)}}{\partial y^{(\mathrm{NIP})}}\right] & \frac{1}{\sigma_p}\left[\frac{\partial p_y^{(\xi)}}{\partial z^{(\mathrm{NIP})}}\right] & \frac{1}{\sigma_p}\left[\frac{\partial p_y^{(\xi)}}{\partial e_x^{(\mathrm{NIP})}}\right] & \frac{1}{\sigma_p}\left[\frac{\partial p_y^{(\xi)}}{\partial e_y^{(\mathrm{NIP})}}\right] & \frac{1}{\sigma_p}\left[\frac{\partial p_y^{(\xi)}}{\partial v}\right] \\[4pt]
\frac{1}{\sigma_\xi}\left[\frac{\partial \xi_x}{\partial x^{(\mathrm{NIP})}}\right] & \frac{1}{\sigma_\xi}\left[\frac{\partial \xi_x}{\partial y^{(\mathrm{NIP})}}\right] & \frac{1}{\sigma_\xi}\left[\frac{\partial \xi_x}{\partial z^{(\mathrm{NIP})}}\right] & \frac{1}{\sigma_\xi}\left[\frac{\partial \xi_x}{\partial e_x^{(\mathrm{NIP})}}\right] & \frac{1}{\sigma_\xi}\left[\frac{\partial \xi_x}{\partial e_y^{(\mathrm{NIP})}}\right] & \frac{1}{\sigma_\xi}\left[\frac{\partial \xi_x}{\partial v}\right] \\[4pt]
\frac{1}{\sigma_\xi}\left[\frac{\partial \xi_y}{\partial x^{(\mathrm{NIP})}}\right] & \frac{1}{\sigma_\xi}\left[\frac{\partial \xi_y}{\partial y^{(\mathrm{NIP})}}\right] & \frac{1}{\sigma_\xi}\left[\frac{\partial \xi_y}{\partial z^{(\mathrm{NIP})}}\right] & \frac{1}{\sigma_\xi}\left[\frac{\partial \xi_y}{\partial e_x^{(\mathrm{NIP})}}\right] & \frac{1}{\sigma_\xi}\left[\frac{\partial \xi_y}{\partial e_y^{(\mathrm{NIP})}}\right] & \frac{1}{\sigma_\xi}\left[\frac{\partial \xi_y}{\partial v}\right] \\[6pt]
\hline
& & \left[\,0_v\,\right] & & & \frac{1}{\sigma_v}\left[\frac{\partial v^{(\mathrm{constr})}}{\partial v}\right] \\[4pt]
& & \left[\,0_{v_q}\,\right] & & & \frac{1}{\sigma_{v_q}}\left[\frac{\partial (\nabla_q v)}{\partial v}\right] \\[4pt]
& & \left[\,0_B\,\right] & & & \left[\,B\,\right]
\end{pmatrix}
. 
$$

$$\tag{5.50}$$

Analogously to the 1D and 2D cases, the submatrices in the upper left part of $\hat{\mathbf{F}}$ are $n_{\mathrm{data}} \times n_{\mathrm{data}}$ diagonal matrices containing the Fréchet derivatives with respect to the NIP model parameters, while the submatrices in the upper right part of $\hat{\mathbf{F}}$ contain the corresponding derivatives with respect to the velocity model parameters (Appendix E). They are $n_{\mathrm{data}} \times n_x n_y n_z$ matrices with elements

$$
\left[\frac{\partial \tau_0}{\partial v}\right]_{i,[(j-1)n_y n_z + (k-1)n_z + l]} = \frac{\partial \tau_{0i}}{\partial v_{jkl}}
\tag{5.51}
$$

for the matrix $\left[\frac{\partial \tau_0}{\partial v}\right]$ and analogous expressions for the other matrices involving velocity derivatives. As in the 2D case, the matrices $\left[\frac{\partial v^{(\mathrm{constr})}}{\partial v}\right]$ and $\left[\frac{\partial (\nabla_q v)}{\partial v}\right]$ contain the Fréchet derivatives with respect to the velocity model parameters corresponding to the constraints discussed in Section 4.5. They have the dimensions $n_{\mathrm{vdata}} \times n_x n_y n_z$ and $n_{\mathrm{data}} \times n_x n_y n_z$, respectively. The required Fréchet derivative quantities are derived in Appendix F. The matrix $\left[B\right]$ is an $n_x n_y n_z \times n_x n_y n_z$ upper triangular matrix identical to $\mathbf{B}$ defined in equation (C.14) using $\mathbf{D}''^{(\mathrm{3D})}$. The zero matrices $\left[0_v\right]$, $\left[0_{v_q}\right]$, and $\left[0_B\right]$ are of dimensions $n_{\mathrm{vdata}} \times 5 n_{\mathrm{data}}$, $n_{\mathrm{data}} \times 5 n_{\mathrm{data}}$, and $n_x n_y n_z \times 5 n_{\mathrm{data}}$, respectively. The weighting factors $\sigma_\tau$, $\sigma_M$, $\sigma_p$, $\sigma_\xi$, $\sigma_v$, and $\sigma_{v_q}$ have already been discussed in Section 5.2.

The matrix equation obtained by substituting $\hat{\mathbf{F}}$, equation (5.50), and a correspondingly defined vector $\Delta \hat{\mathbf{d}}$ into equation (4.21) is, like in the 2D case, solved numerically for $\Delta \mathbf{m}$ with the LSQR algorithm during each nonlinear iteration.

## 5.3.4 Synthetic test example

To demonstrate the 3D tomographic inversion with kinematic wavefield attributes, it is applied to input data that have been directly obtained by dynamic ray tracing along normal rays in a model defined by parameters of the form of equation (5.38).

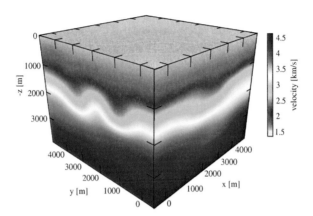

Figure 5.14: 3D synthetic data example. Laterally inhomogeneous velocity model defined by $9 \times 9 \times 9 = 729$ B-spline coefficients on a grid with a 500 m knot spacing in the horizontal and a 400 m knot spacing in the vertical direction.

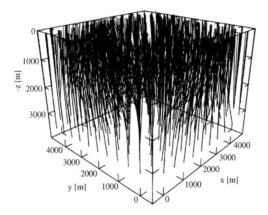

Figure 5.15: 3D synthetic data example. 1008 normal rays corresponding to NIP locations distributed along iso-velocity surfaces in the velocity model displayed in Figure 5.14. The ray starting directions at the respective NIPs point in the direction of the negative local velocity gradient. This simulates the situation that reflectors follow the velocity structure (or vice versa). The input data displayed in Figure 5.16 have been calculated by dynamic ray tracing along these rays.

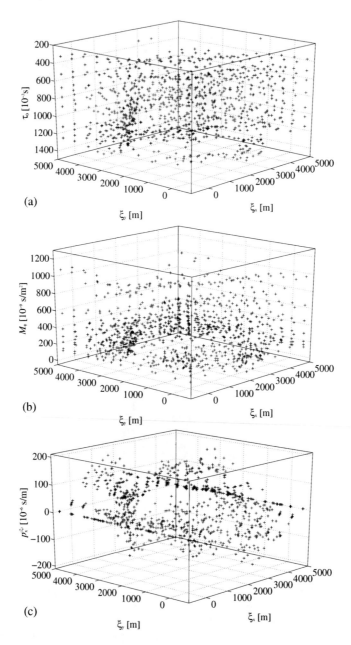

Figure 5.16: 3D synthetic data example. Selected components of the data modeled along the normal rays displayed in Figure 5.15. (a) $\tau_0$, (b) $M_{\mathrm{NIP}}^{(\xi)}$, and (c) $p_x^{(\xi)}$, plotted as a function of the data components $\xi_x$ and $\xi_y$. The data component $p_y^{(\xi)}$ is not displayed.

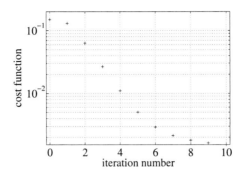

Figure 5.17: 3D synthetic data example. Value of the cost function (4.16) as a function of iteration number. The process converges within 10 iterations.

**Generating test data**

A 3D laterally inhomogeneous velocity model, Figure 5.14, is defined by specifying the values of the B-spline coefficients $v_{jkl}$ on a grid of, in this example, $n_x \times n_y \times n_z = 9 \times 9 \times 9$ knot locations with a constant horizontal spacing in $x$ and $y$ of 500 m and a vertical spacing of 400 m. Depth slices through that model are displayed in the left column of Figure 5.19, while vertical sections through the model at a number of $x$-locations are shown in the left column of Figure 5.20. In this velocity model, a total of 1008 NIP model points are placed. As in the 2D example described in Section 5.2, the NIPs are placed in the velocity model along iso-velocity surfaces in such a way that the corresponding local reflector orientations defined by $(e_x, e_y)^{(\mathrm{NIP})}$ follow these iso-velocity surfaces. For each of these NIP model points, the quantities $(\tau_0, M_\phi^{(\xi)}, p_x^{(\xi)}, p_y^{(\xi)}, \xi_x, \xi_y)$ constituting a data point are computed by performing dynamic ray tracing along the normal ray defined by the respective NIP model parameters. The 1008 normal rays are plotted in Figure 5.15. Figures 5.16 show the data components $\tau_0$, $M_\phi^{(\xi)}$, and $p_x^{(\xi)}$, plotted as a function of the emergence location $(\xi_x, \xi_y)$. All data components (5.37) of the 1008 data points serve as input for the tomographic inversion with the aim of reconstructing the true velocity model and NIP locations.

**Inversion**

The initial velocity model for the tomographic inversion is in this example defined by a near-surface velocity of 2000 m/s and a constant vertical velocity gradient of 0.5 s$^{-1}$. As described in Section 4.6, initial NIP model parameters are found for each ray by ray tracing from the measurement surface into the model until the traveltime $\tau_0$ is used up. Similarly to the 2D example discussed in the previous section, the constraint of minimum local velocity gradient in the plane tangent to the reflector, equation (5.40), is applied with a weighting factor of $\sigma_{v_q} = 10$.

The tomographic inversion is performed with 10 nonlinear iterations, again using equation (5.20) to control the decrease of the overall regularization from iteration to iteration. The decrease of

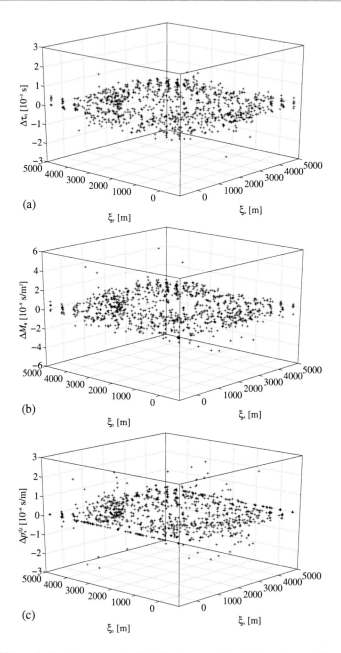

Figure 5.18: 3D synthetic data example. Residual errors after 10 iterations of the selected data components displayed in Figure 5.16. (a) $\Delta\tau_0$, (b) $\Delta M_{\mathrm{NIP}}^{(\xi)}$, and (c) $\Delta p_x^{(\xi)}$. Residual errors of $p_y^{(\xi)}$, $\xi_x$, and $\xi_y$ are not displayed.

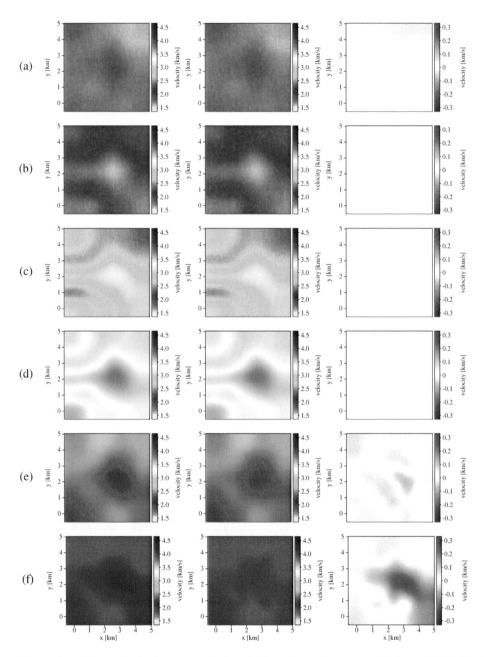

Figure 5.19: 3D synthetic data example. Depth slices through the true velocity model (left), the inversion result (center) and the difference between the true model and the inversion result (right) at (a) z=-500 m, (b) z=-1000 m, (c) z=-1500 m, (d) z=-2000 m,(e) z=-2500 m, and (f) z=-3000 m.

95

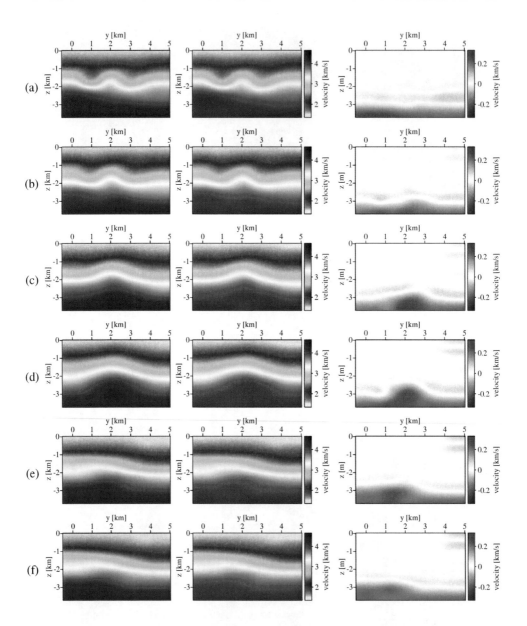

Figure 5.20: 3D synthetic data example. Vertical slices through the true velocity model (left), the inversion result (center) and the difference between the true model and the inversion result (right) at (a) x=0 m, (b) x=1000 m, (c) x=2000 m, (d) x=3000 m,(e) x=4000 m, and (f) x=5000 m.

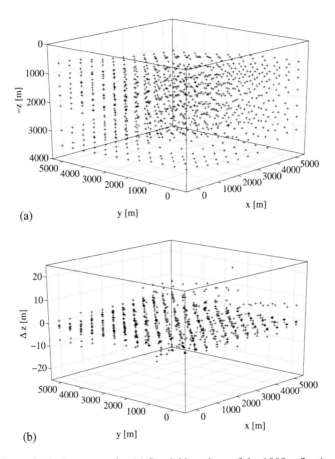

(a)

(b)

Figure 5.21: 3D synthetic data example. (a) Spatial locations of the 1008 reflection points (NIPs) placed in the 3D velocity model. (b) Residual depth errors after 10 iterations of the reconstructed reflection points relative to the true reflection points displayed in (a).

the cost function (4.16), plotted as a function of iteration number in Figure 5.17, indicates a good convergence of the process.

**Results**

Figures 5.18a, b, and c show the residual misfit in the data components of Figure 5.16 after 10 iterations. The inversion result itself consists of the reconstructed velocity model and NIP model parameters corresponding to the input data. Depth slices through the reconstructed velocity model for the depths of 500 m, 1000 m, 1500 m, 2000 m, 2500 m, and 3000 m are displayed in the center column of Figure 5.19, while the center column of Figure 5.20 shows vertical sections through the reconstructed model at a number of different $x$-locations. The difference volume calculated by subtracting the reconstructed velocity model from the true velocity model is displayed in the form of depth slices and vertical sections in the right columns of Figures 5.19 and 5.20, respectively. Differences are mainly visible in the lowermost part of the model, which is not constrained by the data as it is not covered by rays. Apart from that, the velocity model has been well reconstructed, especially considering the low number of data points used for the inversion. The good model reconstruction is also confirmed by evaluating the reconstruction of the NIP locations. The true NIP locations are plotted in Figure 5.21a, while Figure 5.21b shows the depth errors of the reconstructed NIPs with respect to the depths of the true NIPs. The depth error remains below 10 m for almost all NIPs, again indicating that the model has been well reconstructed by the tomographic inversion.

# Chapter 6

# Applications

In Chapters 4 and 5 the tomographic inversion based on NIP wave kinematic wavefield attributes has been introduced and demonstrated on exact synthetic input data. In practice, however, the input data for the tomographic inversion are obtained from seismic data by performing the CRS stack (Chapter 3) and subsequently extracting the required kinematic wavefield attributes at a number of zero-offset pick locations from the CRS stack results.

In this chapter, practical aspects of the new inversion method are discussed with special emphasis on issues related to the extraction of reliable input data (Section 6.1). The entire inversion process, starting with the seismic prestack data and leading to the final velocity model that can be used for prestack and poststack depth migration, is then demonstrated on a synthetic 2D seismic dataset (Section 6.2) and a real 2D land dataset (Section 6.3).

## 6.1 Practical aspects

The tomographic inversion method introduced in Chapters 4 and 5 is based on the idea that second-order traveltime information is sufficient for reliably determining velocity models for depth imaging even in laterally inhomogeneous media, as long as lateral variations are not too strong. As already discussed in Chapter 4, this use of second-order traveltime approximations leads to considerable simplifications in the process of extracting kinematic information from the seismic prestack data. All required information is contained in the kinematic wavefield attributes automatically obtained from the prestack data with the CRS stack. Kinematic wavefield attributes are, however, determined for every zero-offset sample, irrespective of whether it is located on an actual reflection event or not. The zero-offset sample locations for which the associated kinematic wavefield attributes are to be used as input for the tomographic inversion therefore need to be picked to ensure that only attributes related to actual reflection events contribute. A great advantage of the specific model parametrization used in the tomographic inversion as introduced in Chapter 4 lies in the fact that all pick locations can be considered independently of each other and do not need to follow interpreted horizons in the seismic data. Locally coherent events in the simulated zero-offset section/volume are sufficient, which significantly simplifies the picking process.

However, the extensive use of approximations also leads to certain limitations in the applicability of the method. Obviously, second-order traveltime approximations have only a limited range of validity, especially in complex media. This needs to be taken into account during the application of the CRS stack in order to obtain reliable kinematic wavefield attributes. Also, the use of dynamic ray tracing along single normal rays to model second-order traveltimes of emerging NIP waves demands a certain degree of smoothness of the model in order to ensure that the resulting kinematic wavefieldd attributes can be related to those which have been extracted from the seismic data with the CRS stack. As with all traveltime inversion methods, the picking process itself may lead to errors which need to be taken into account. In the following, practical aspects to be considered in the different steps of the inversion process will be discussed.

## 6.1.1 CRS stack

The aim of the CRS stack process in the context of velocity model estimation is to provide reliable traveltime information in the form of the kinematic wavefield attributes, given by equation (3.8) for the 2D case or by equation (3.12) for the 3D case, respectively, for each zero-offset sample. In fact, the attributes in equations (3.8) and (3.12) are local quantities associated strictly with the considered zero-offset location and traveltime. They are treated as such when they are used as input data for the tomographic inversion. However, when they are determined from the seismic prestack data using a traveltime equation of the form of equation (3.7) or (3.11), respectively, a finite search aperture in the midpoint and offset directions is used. This leads to inaccuracies in the determined kinematic wavefield attributes, the severity of which depends, among other factors, on the deviation of the actual reflection traveltimes from the second-order approximations (3.7) or (3.11), and, thus, on the complexity of the subsurface. Even in 1D layered media, traveltimes are not strictly hyperbolic (Section 3.2). In order to minimize the effects of deviations of reflection events from the approximations (3.7) or (3.11), the search apertures in midpoint $\xi_m$ and half-offset $h$ would have to be chosen as small as possible. On the other hand, it is well known from conventional stacking velocity analysis that in order to obtain reliable moveout parameters (kinematic wavefield attributes), the search aperture needs to be sufficiently large. This is especially true for deep reflectors, where the resolution in the offset direction is degraded due to a small overall moveout. A sufficiently large aperture is also required to reduce the effect of random noise in the seismic data by allowing a large number of traces to contribute.

A list of different factors affecting the accuracy and resolution of stacking velocities is given by Yilmaz (2001). The items listed there are equally valid for the determination of kinematic wavefield attributes with the CRS stack. They include

- spread length (offset and midpoint aperture),

- stacking fold,

- S/N ratio,

- muting,

- time gate length (for the coherence analysis),

- velocity (or attribute) sampling,

- choice of coherence measure,

- true departures from hyperbolic moveout, and

- bandwidth of the data.

An especially crucial factor is the deviation of reflection traveltimes from hyperbolic moveout. If this deviation is too large, indicating that the subsurface complexity is too severe, a reliable determination of the kinematic wavefield attributes becomes impossible and the tomographic inversion based on these attributes is not applicable. In order to examine if this is the case, it is sufficient to visually check a number of CMP gathers to see if, or up to what maximum offset, reflection events look approximately hyperbolic. Practical experience suggests that for target depths up to 3 or 4 km, a maximum offset of 2 to 2.5 km is sufficient. Due to the fact that in the inversion, a smooth model description is used and the model is constrained by the data in a least-squares sense, a certain degree of deviation from hyperbolic moveout in the considered aperture, leading to a certain level of data error, can be tolerated. The stability of the inversion process with respect to data errors has been demonstrated in Section 5.2. Also, the averaging of velocity inhomogeneities during the dynamic ray tracing, as described in Section 5.2, may be used to account for departures of actual reflection traveltimes from a hyperbolic shape.

### 6.1.2 Smoothing of attributes

During the CRS stack, optimum kinematic wavefield attributes are determined independently for each zero-offset sample. Although this has a number of advantages like providing high-resolution attribute information and avoiding NMO stretch (Mann and Höcht, 2003), it may lead to a certain unphysical fluctuation of the determined attributes along a reflection event. As already pointed out in Section 3.5, this fluctuation can be corrected for by applying an event-consistent smoothing. Such a smoothing procedure can be implemented in a simple way as described in Appendix G. Apart from the positive effect on the stack result itself (Figure G.1), the event-consistent attribute smoothing helps to avoid an unphysical scatter of data values used in the tomographic inversion, which could not be explained with a smooth velocity model and would therefore degrade the convergence of the inversion process. The usefulness of the smoothing of attribute sections or volumes depends on the noise level in the seismic data. The process does not solve the problem of non-hyperbolic moveout.

### 6.1.3 Picking

Once the CRS stack has been performed and the (possibly smoothed) kinematic wavefield attribute sections/volumes are available, the picking of zero-offset points can be performed. In the following, the term "section" will be used to signify either a data section or a data volume, depending on whether the 2D or the 3D case is considered. As noted above, the pick locations are independent of each other and do not need to follow interpreted horizons. They do, however, need to be located on primary reflection events. The picking can be performed directly in the simulated zero-offset section or in the associated CRS coherence section. The simulated zero-offset section has the advantage that it allows to perform a somewhat interpretative picking and possibly identify and avoid multiple reflections which would be much more difficult to identify in the prestack data. Using the CRS coherence section, on the other hand, allows to directly identify events with a high coherence and, thus, reliable wavefield attributes. After the picking has been performed, the kinematic wavefield attributes at the picked zero-offset sample locations are automatically extracted from the corresponding attribute sections.

If in a given seismic dataset multiples and reverberations are not a problem (or have been removed), the picking of input data points for the tomographic inversion can be automated. For that purpose, it is sufficient to simply automatically pick maxima in the (possibly filtered) CRS coherence section and apply additional rules like a coherence threshold, a minimum separation in space and time of pick locations and some constraints on the allowed values for the kinematic wavefield attributes.

The required spatial and temporal separation of picks depends on the chosen velocity model parametrization, that is, on the horizontal and vertical B-spline knot spacing. In the horizontal direction, there should be at least two (preferably more) picks per reflection event per horizontal knot interval. As a general rule, the number of data should exceed the number of model parameters otherwise the problem is underdetermined and the regularization weight needs to be increased, leading to an unrealistically smooth model. A larger number of data points leads to an improved stability of the inversion process, with a roughly linear increase in computation time. However, the $L_2$ norm, equation (4.9), used as a measure of data misfit in the tomographic inversion is very sensitive to unreliable data which lead to large data misfits. To avoid the negative effects of such outliers on the inversion process, data points to be used in the inversion should be carefully selected, based on their expected reliability, rather than using large numbers of possibly unreliable data points.

In the temporal direction, the minimum separation between picks should exceed the reflection signal length to ensure that two picks at the same spatial location actually correspond to separate reflection events. Problems can occur where reverberations are present and different parallel reflection events cannot be properly separated.

### 6.1.4  Editing of picks

After the picking has been performed and the different required wavefield attributes have been extracted from the CRS attribute sections, the resulting set of data points has to be checked with respect to its reliability and possibly needs to be edited. During this editing step, unreliable data points and points likely to be related to multiple reflections are identified and removed. Such unreliable data can be identified by plotting the different data components as a function of their spatial coordinate. Theoretically, the kinematic wavefield attributes should vary smoothly along a reflection event. Therefore, the data components of points corresponding to a common reflection

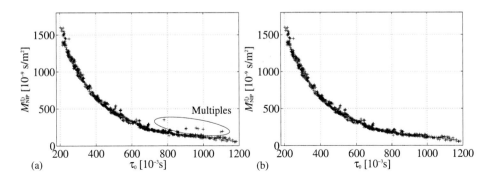

Figure 6.1: Identification of multiples by plotting $M_{\text{NIP}}^{(\xi)}$ against $\tau_0$. In simple media, multiples have systematically higher values of $M_{\text{NIP}}^{(\xi)}$ than primary reflections of comparable traveltimes. (a) Original picked data from the real data example of Section 6.3. (b) Picked data after removal of multiples.

event should exhibit a certain continuity. This can be well observed in the real data example, Figure 6.14. Obvious outliers and data values which fall far from the trend of neighboring values should be removed.

In general, data points with very high values of $M_{\text{NIP}}^{(\xi)}$ (significantly larger than $2000 \cdot 10^{-9}$ s/m$^2$) should be avoided, as these tend to destabilize the inversion process. High values of $M_{\text{NIP}}^{(\xi)}$ occur for shallow reflection events (small traveltimes $t_0$).

An important issue for the tomographic inversion is the removal of data points related to multiple reflections. If subsurface structures are relatively simple, such data points can be identified in a way similar to the conventional velocity discrimination method (e. g., Yilmaz, 2001). According to equations (3.8) and (3.10) (or equations (3.13) and (3.14), respectively), $M_{\text{NIP}}^{(\xi)}$ is related to the normal moveout velocity $v_{\text{NMO}}$ via

$$M_{\text{NIP}}^{(\xi)} = \frac{2}{t_0 \, v_{\text{NMO}}^2} \; . \tag{6.1}$$

The characteristic decrease of $v_{\text{NMO}}$, as normally observed for multiple reflections, therefore translates into a corresponding increase of $M_{\text{NIP}}^{(\xi)}$ at the considered traveltime. If the $M_{\text{NIP}}^{(\xi)}$ values are plotted against $\tau_0 = t_0/2$ (Figure 6.1a) data points associated with multiple reflections show up above and roughly parallel to the main trend and can be eliminated, see Figure 6.1b. In a 3D plot of $M_{\text{NIP}}^{(\xi)}$ values against $\xi_0$ and $\tau_0$, data points suspected to be associated with multiples can directly be related to events in the zero-offset section , which further helps in identifying multiples.

In the case of complex subsurface structures, this method of multiple identification will no longer be reliable and other ways of addressing the problem of multiple reflections need to be found.

### 6.1.5  Model parametrization, regularization, and constraints

Before the tomographic inversion can be started, the velocity model needs to be defined. This includes choosing suitable B-spline knot sequences in the horizontal and vertical directions and specifying the velocity values for the initial model. The proper choice of vertical and horizontal B-spline knot spacings depends on the expected complexity of the subsurface velocity structure. However, the approximations made in the tomographic inversion also need to be taken into account. As already discussed, the assumption that the values of $M_{\text{NIP}}^{(\xi)}$, or $M_{\phi}^{(\xi)}$, determined from the seismic data with the CRS stack can be related to the corresponding quantities modeled by dynamic ray tracing along single normal rays implies a certain smoothness of the velocity model in the horizontal direction. The following quantitative considerations apply to seismic exploration situations with target depths of the order of 2 to 4 km. If an offset aperture of 2 km or more is used during the CRS stack, the horizontal B-spline spacing should be at least 500 m. Although the B-spline knot spacing does not need to be constant (Appendix B), there is usually no practical reason for choosing a variable knot spacing in the horizontal direction. In the vertical direction, however, a variable B-spline knot spacing may very well be useful. It is a common situation that the shallow part of the subsurface is well constrained by a large number of reflection events, while only few reflection events are visible at larger traveltimes. In such situations it is useful to choose a denser vertical knot spacing in the shallow part of the model than in the deeper part. In general, vertical knot spacings with values anywhere between 100 m and 500 m are reasonable.

The easiest way of defining an initial velocity model for the inversion is to specify a constant-vertical-gradient model of the form

$$v(z) = V_0 + az \,, \tag{6.2}$$

where reasonable values for $V_0$ and $a$ should be chosen based on a priori geological information. Alternatively, if more detailed information is already available, it is possible to explicitly assign an initial value to each of the B-spline coefficients $v_{jk}$ or $v_{jkl}$, respectively.

A crucial point in the application of all tomographic inversion methods is the proper choice of the regularization weight. What needs to be considered is the relative weighting of the different terms in equations (C.10) or (C.11), respectively, as well as the overall regularization weight $\varepsilon''$. The relative weighting is performed by specifying values for the factors $\varepsilon_{xx}$, $\varepsilon_{yy}$, $\varepsilon_{zz}$, and $\varepsilon$ (see Appendix C), which may in principle all be chosen independently. In practice, it is, however, usually reasonable to set $\varepsilon_{xx} = \varepsilon_{yy} = \varepsilon_{zz}$, while the value of $\varepsilon$ should be chosen to be significantly smaller than the other factors. The reason for this is discussed in Appendix C. A reasonable initial value for the overall regularization weight $\varepsilon''$ needs to be found interactively. Different choices of $\varepsilon''$ lead to different degrees of smoothness of the final model. While a too large value may force the velocity model to be excessively smooth, hindering the decrease of the data misfit and, thus, the convergence of the process, a too low value may lead to instabilities in the inversion.

As discussed in Section 4.6 and in Section 5.1, in order to enhance the convergence behavior of the inversion process, the overall regularization weight $\varepsilon''$ is decreased from iteration to iteration by using a relation like equation (5.20). In practice, the initial overall regularization factor $\varepsilon''$ can be combined with the relative regularization weights $\varepsilon_{xx}$, $\varepsilon_{yy}$, $\varepsilon_{zz}$, and $\varepsilon$, so that only these need to be specified (implying an initial value of $\varepsilon''$ of 1).

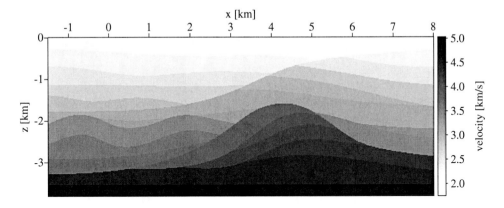

Figure 6.2: Synthetic data example. Blocky P-wave velocity distribution used to generate a synthetic multicoverage seismic dataset by ray-tracing modeling. The velocity discontinuities act as reflectors. The corresponding S-wave velocity and density models are not displayed.

In general, for properly choosing a regularization weight, the degree of model detail that can be expected to be resolved with the given data needs to be considered. Oscillations of the velocity model in between reflection points or in areas, where very few or no NIP locations related to data points are present have to be regarded as artificial structure, indicating that a too low regularization weight has been used. If parts of the model are expected to be well constrained by the data, while other parts are not constrained at all, the spatially variable smoothness constraint discussed in Section 4.5 can be applied. All of the constraints treated in Section 4.5 are applied in the 2D real data example presented in Section 6.3.

## 6.2   2D synthetic data example

In order to demonstrate the complete process of constructing a velocity model for depth migration from seismic prestack data using the attribute-based tomographic inversion, it is first applied to a synthetic dataset. This allows to evaluate the performance of the inversion process under controlled conditions.

### 6.2.1   The synthetic seismic dataset

To start with, a 2D prestack seismic dataset is generated by ray-tracing modeling in the laterally inhomogeneous, blocky velocity model shown in Figure 6.2. P-wave velocities in the model range from 2000 m/s to 5000 m/s. The corresponding S-wave velocity and density models are not displayed, as only primary P-wave reflections are modeled and only the kinematics of reflection

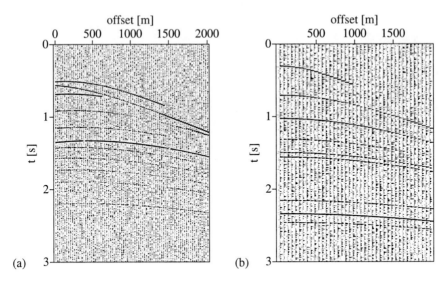

Figure 6.3: Synthetic data example. (a) Sample shot gather for the shot location $x = 5000$ m. (b) Sample CMP gather for the CMP location $x = -500$ m. Only primary P-wave reflection events are modeled, diffractions are not considered.

events are relevant for the application of the tomography (as long as all attributes can be reliably determined). Note that no diffraction events are modeled.

For the modeling, shots are placed at the measurement surface ($z = 0$) every 25 m and reflections are recorded at receivers with a spacing of 25 m and an offset range of 0 to 2000 m in a marine-type acquisition geometry. A total of 497 shots, each recorded at 81 receivers, are modeled, with shot coordinates ranging from $x = -2000$ m to $x = 10400$ m. As source signal, a 30 Hz zero-phase Ricker wavelet is used. The sampling interval is 4 ms and the maximum recording time is 3 s. Band-limited noise is added to the resulting seismic prestack dataset. A sample shot gather (shot location: $x = 5000$ m) is displayed in Figure 6.3a, while Figure 6.3b shows a sample CMP gather (CMP location: $x = -500$ m).

The objective is to use the tomographic inversion based on kinematic wavefield attributes to derive a velocity model from this dataset that is kinematically equivalent to the true model (Figure 6.2) for all reflection events in the data. The resulting smooth velocity model can then be used to perform prestack and poststack depth migration to obtain a correct structural image of reflectors in the model. Note that the model obtained as a result of the tomographic inversion cannot be identical to the true model, as it is described by smooth basis functions, while the true model is blocky. However, the inversion result should be kinematically consistent with the data, or even kinematically equivalent to the true model for all reflection events in the data.

### 6.2.2 CRS stack

As a first step in the inversion procedure, the CRS stack is applied to the prestack data. The search aperture in the offset direction used during this process is defined to range from 1000 m offset at 0.2 s to 2000 m offset at 1.5 s, while the full midpoint search aperture ranges from 200 m at traveltime zero to a maximum value of 300 m. The CRS stack is performed using the implementation of Mann (2002) based on equation (3.9) in terms of emergence angle $\alpha$ and radii of wavefront curvature $R_{NIP}$ and $R_N$. For the near-surface velocity a value of $v_0 = 2000$ m/s is assumed. The applied search strategy is described in Section 3.5. It involves three separate one-parameter searches and a subsequent three-parameter local optimization. Conflicting dips are not considered in this example.

The results of the CRS stack process that are relevant for the tomographic inversion are displayed in Figures 6.4 and 6.5. These are the CRS stacked simulated zero-offset section, Figure 6.4a, the CRS coherence section, Figure 6.4b, the emergence angle section, Figure 6.5a, and the section containing the NIP wave radius of wavefront curvature, Figure 6.5b. The coherence section contains a semblance value for each zero-offset sample, indicating where reliable attributes have been determined. The coherence on all reflection events is high enough to warrant reliable attributes, while in between the reflection events the coherence is virtually zero. Correspondingly, attributes not associated with a reflection event have no physical significance. The emergence angles in Figure 6.5a and the $R_{NIP}$ values in Figure 6.5b behave as expected. The angles are directly linked to the local reflector dips in the zero-offset section, while the values of $R_{NIP}$ increase systematically with traveltime. Note the high $R_{NIP}$ values in the central part of the lowermost reflection event. They correspond to waves propagating through the most complex part of the model.

In this synthetic example, no smoothing of the kinematic wavefield attribute sections is required. Additional results of the CRS stack process, including a normal wave curvature section and a stacking-velocity section are not displayed as they are not needed for the tomographic inversion.

### 6.2.3 Tomographic inversion

In the CRS stack section (Figure 6.4a) a total of 500 zero-offset points are picked and the associated kinematic wavefield attributes are automatically extracted from the corresponding sections (Figures 6.5a and b). The resulting data components $\tau_0$, $M_{NIP}^{(\xi)}$, and $p^{(\xi)}$, calculated from the extracted values of $R_{NIP}$ and $\alpha$ via equation (3.8), are displayed in Figures 6.6a, c, and e as a function of the emergence location $\xi_0$ (the fourth data component). As demanded in Section 6.1, the different input data components appear to be varying smoothly along reflection events and can therefore be regarded as reliable. These 500 data points are used as input for the tomographic inversion.

The velocity model to be determined is defined on a grid of $n_x \times n_z = 17 \times 13$ B-spline knots with a constant horizontal knot spacing of 500 m and a constant vertical knot spacing of 300 m. Knot locations range from $x = -1000$ m to $x = 7000$ m horizontally and from the measurement surface ($z = 0$) to a depth of 3600 m vertically. For the initial model, a near-surface velocity of 2000 m/s and a constant vertical velocity gradient of 0.67 s$^{-1}$ is used.

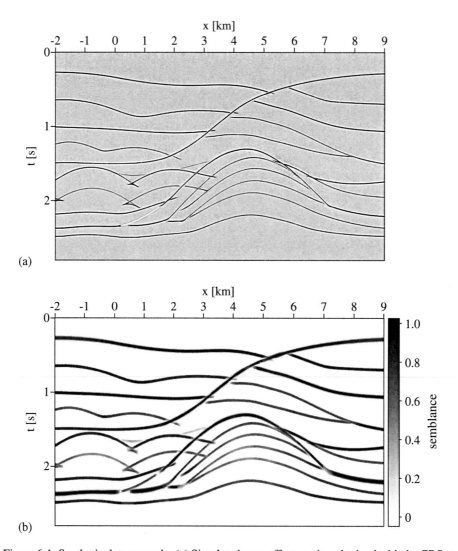

Figure 6.4: Synthetic data example. (a) Simulated zero-offset section obtained with the CRS stack, (b) CRS coherence section, containing the semblance values, equation (3.5), obtained along the optimum CRS operators.

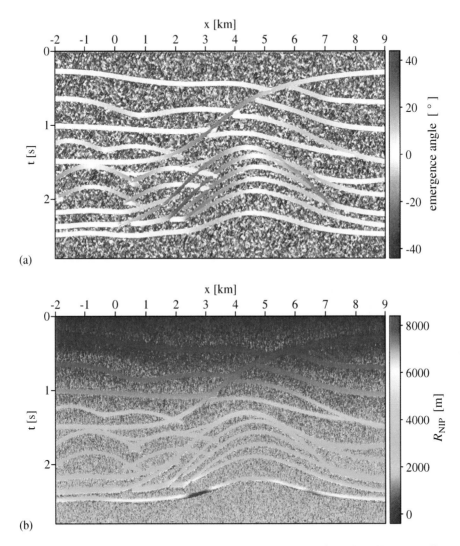

Figure 6.5: Synthetic data example. (a) Emergence angle $\alpha$ section. (b) NIP wave radius $R_{\mathrm{NIP}}$ section.

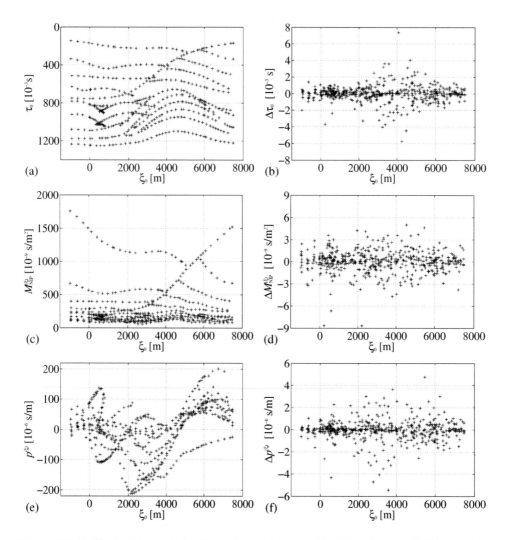

Figure 6.6: Synthetic data example. Input data and final residual data errors. (a) Values of the data component $\tau_0$ used in the tomographic inversion. (b) Residual errors in $\tau_0$ after 12 iterations. (c) Values of the data component $M_{NIP}^{(\xi)}$ used in the tomographic inversion. (d) Residual errors in $M_{NIP}^{(\xi)}$ after 12 iterations. (e) Values of the data component $p^{(\xi)}$ used in the tomographic inversion. (f) Residual errors in $p^{(\xi)}$ after 12 iterations.

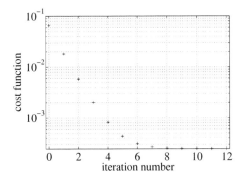

Figure 6.7: Synthetic data example. Value of the cost function (4.16) as a function of iteration number. The process effectively converges within 8 iterations.

The regularization (Appendix C) is applied with $\varepsilon_{xx} = \varepsilon_{zz}$ in equation (C.10) and an overall regularization weight that decreases from iteration to iteration according to relation (5.20). However, none of the additional model constraints described in Section 4.5 are applied. For the data weights $\sigma_\tau$, $\sigma_M$, $\sigma_p$, and $\sigma_\xi$, the values suggested in Section 5.2 are used. Also, during the modeling and calculation of Fréchet derivatives, the higher spatial velocity derivatives are averaged around the central ray as described in Section 5.2, with a maximum averaging width of 500 m at $z = 0$.

With these parameters, a total of 12 nonlinear inversion iterations are performed. However, as can be observed in Figure 6.7, where the value of the cost function is plotted as a function of iteration number, the process already effectively converges after 8 iterations. The final residual misfits in the data components $\tau_0$, $M_{\mathrm{NIP}}^{(\xi)}$, and $p^{(\xi)}$ are displayed in Figures 6.6b, d, and f. All data misfits have been well reduced and lie in the order of the expected input data error (Section 5.2). The inversion result after 12 iterations is displayed in Figure 6.8. Figure 6.8a shows the smooth velocity model obtained with the tomographic inversion, along with the 500 normal rays corresponding to the final NIP model parameters $(x, z, \theta)^{(\mathrm{NIP})}$. These final NIP parameters, represented by local reflector elements, or dip bars, are plotted into the reconstructed velocity model in Figure 6.8b. The local reflector elements follow the true subsurface structure very well. This is confirmed in Figure 6.8c, where the reconstructed reflector elements are plotted into the true blocky velocity model. All reflector elements fall almost exactly onto the true velocity discontinuities, indicating that the obtained smooth velocity model is indeed kinematically equivalent to the true model for all considered reflection events.

### 6.2.4 Depth migration

The smooth velocity model obtained with the tomographic inversion can now be used to perform a prestack or poststack depth migration. For that purpose, a Kirchhoff migration algorithm (Hertweck, 2004) based on eikonal traveltimes will here be used. In Figures 6.9 and 6.10 the result

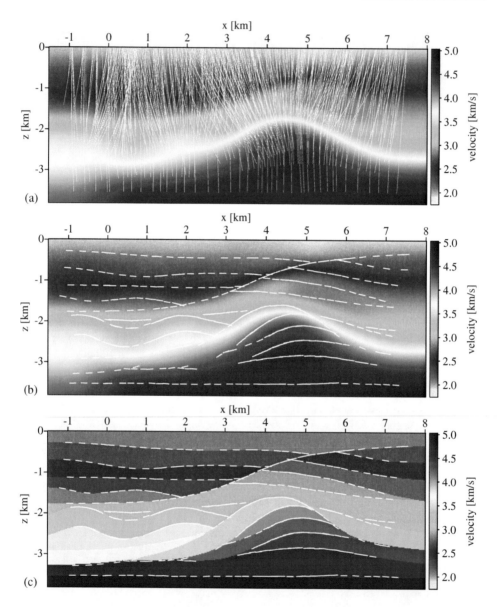

Figure 6.8: Synthetic data example. Inversion result after 12 iterations. (a) Reconstructed smooth velocity model and normal rays corresponding to the final NIP model parameters. (b) Reconstructed velocity model and local reflector elements representing the final NIP model parameters obtained during the tomographic inversion. (c) The reconstructed reflector elements of (b) plotted into the true, blocky velocity model. The reflector elements fall almost exactly onto the discontinuities of the true model, indicating that the reconstructed smooth model is kinematically correct.

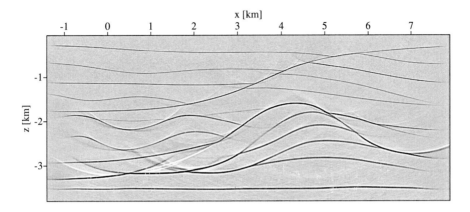

Figure 6.9: Synthetic data example. Kirchhoff prestack depth migration with the smooth velocity model obtained by tomographic inversion. A stack of common-offset migrations for an offset range between 0 and 2000 m is shown. Artefacts are due to missing diffraction events in the modeled prestack data.

of a prestack depth migration using the smooth tomographic velocity model is displayed. Figure 6.9 shows a stack of all common-offset migrations between 0 and 2000 m offset (muted as in Figure 6.10). As expected from the results of Figure 6.8, all reflectors in the migrated image are correctly positioned. In particular, the lowermost reflector is almost perfectly horizontal. The pull-up effect of reflection events in the seismic data due to the high-velocity dome, as is visible, for example, in the CRS stack results in Figure 6.4 and 6.5, has been correctly undone. The migration artefacts are due to the absence of diffraction events in the original seismic prestack data. A poststack depth migration of the CRS stack section (Figure 6.4a) yields a similar image and is not displayed here. Figure 6.10 shows a number of CIGs at regularly spaced image locations with a separation of 500 m. Each CIG represents the migration result at the respective image location as a function of offset. The offsets displayed in the CIGs in Figure 6.10 range from 0 to 2000 m. At shallow depths and large offsets, a mute has been applied to remove events with excessive wavelet stretch. As discussed in Section 1, a velocity model is consistent with the seismic data if the results of prestack depth migration are kinematically independent of offset, that is, if events in the CIGs are flat. This is clearly the case in the CIGs displayed in Figure 6.10, which again confirms that the model obtained with the tomographic inversion is kinematically correct.

The synthetic data example presented in this section shows that the concept of using second-order traveltime approximations for the construction of smooth, laterally inhomogeneous velocity model is indeed applicable, even if the true subsurface velocity distribution is not smooth. There is, however, a limit to the spatial wavelength and magnitude of lateral velocity variations that can be handled. As discussed in Section 6.1, this limit is related to the size of the offset aperture that is required to obtain reliable kinematic wavefield attributes. Within this aperture, reflection events need to be roughly hyperbolic.

113

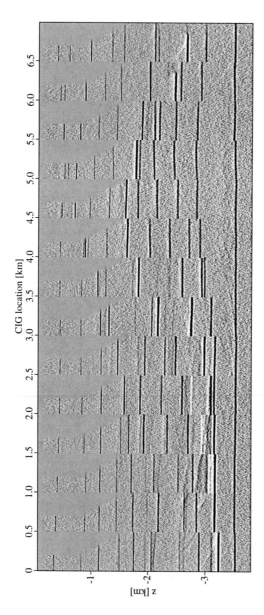

Figure 6.10: Synthetic data example. Selected common-image gathers (CIGs) resulting from Kirchhoff prestack depth migration using the smooth velocity model obtained with the tomographic inversion are shown. The offset in each CIG ranges from 0 to 2000 m. Almost all events in the CIGs are flat, indicating that the model is kinematically correct.

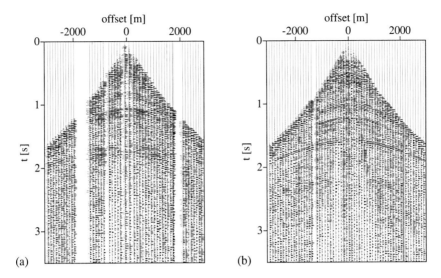

Figure 6.11: Real data example. Two selected CMP gathers from the HotRock dataset. The displayed offset range is limited to 3000 m.

## 6.3    2D real data example

The complete attribute-based tomographic velocity model estimation process, demonstrated on synthetic seismic data in Section 6.2, will now be applied to a real 2D seismic land dataset.

### 6.3.1    The seismic dataset

The seismic data presented here were acquired in 2003 in the Oberrheingraben near Karlsruhe for HotRock EWK Offenbach/Pfalz GmbH with the aim of investigating geological structures at the site of a planned geothermal power plant. In particular, the goal was to obtain an image of the production horizon to be drilled, a carbonate formation with a thickness of about 110 m at roughly 2.5 km depth, and to delineate faults which play a significant role for the transport of geothermal water.

The seismic acquisition was carried out by Deutsche Montan Technologie GmbH (DMT) along two 12 km long parallel profiles (a North and a South profile, roughly 3 km apart), one of which will be used here. For each of the two profiles the data were recorded with a fixed spread of receivers covering the entire length of the respective profile. Both, the receiver group as well as the source spacing were 50 m, leading to a CMP separation of 25 m. The source energy was provided by three seismic vibrators in the form of a linear sweep from 12 to 100 Hz and was recorded with a sampling interval of 2 ms. The entire preprocessing, including amplitude corrections, trace editing, static corrections, mute, deconvolution and filtering, was performed by DMT. Two CMP gathers from the North profile, with maximum offsets restricted to 3 km, are displayed in Figures 6.11a and b.

The preprocessed seismic prestack data of the North profile, provided by DMT, are directly used as the starting point for the determination of a velocity model with the attribute-based tomographic inversion. For practical reasons, a new coordinate system with the $x$-axis along the profile direction is used in the following.

## 6.3.2 CRS stack

As a first step of the inversion process, the CRS stack is performed. For the reasons discussed in Section 6.1, the offset aperture for the CRS stack is restricted to linearly range from 200 m offset at 0.2 s traveltime to 2000 m offset at 1.2 s traveltime. The full midpoint aperture is defined to range from 300 m at zero traveltime to a maximum of 500 m at later times. Again, the CRS stack is performed using the implementation of Mann (2002) in terms of emergence angle $\alpha$ and radii of wavefront curvature $R_{NIP}$ and $R_N$, equation (3.9). For the near-surface velocity, a value of $v_0 = 1700$ m/s is used. The CRS attribute determination is performed in three separate one-parameter searches, as described in Section 3.5. The $\alpha$ and $R_{NIP}$ sections are then smoothed using the smoothing algorithm of Appendix G. As shown in Figure G.1, the stack result is significantly improved by this smoothing process. The smoothed kinematic wavefield attributes then serve as input for a final local three-parameter search, to obtain optimum kinematic wavefield attributes for each zero-offset sample.

The final CRS stack results relevant for the tomographic inversion are displayed in Figures 6.12 and 6.13. These are the simulated zero-offset section itself, the CRS coherence section, the emergence angle $\alpha$ section, and the NIP wave radius $R_{NIP}$ section. Note that in the $\alpha$ and $R_{NIP}$ sections, only attributes corresponding to zero-offset samples with a sufficiently high CRS coherence are displayed. Samples with semblance values below 0.01 are masked. Low coherence can mainly be observed between CMP locations 500 and 550, where the recorded wavefield is disrupted by the presence of faults in the subsurface, and below the strong reflection event at about 1.5 s.

## 6.3.3 Tomographic inversion

Once the CRS stack results are available, the input data for the tomographic inversion can be obtained. For that purpose, about 800 zero-offset points are picked in the CRS stack section (Figure 6.12a). The corresponding kinematic wavefield attributes are then automatically extracted from the attribute sections (Figures 6.13a and b). After eliminating outliers and data points likely to be related to multiples (Section 6.1), a total of 793 data points remain. The resulting data components $\tau_0$, $M_{NIP}^{(\xi)}$, and $p^{(\xi)}$ are plotted as a function of the fourth data component $\xi_0$ in Figures 6.14a, c, and e. These data points are used as input for the tomographic inversion.

The velocity model to be determined is defined on a grid of $n_x \times n_z = 21 \times 16$ B-spline knots with a constant horizontal spacing of 500 m and a constant vertical spacing of 200 m. Knot locations associated with B-spline coefficients range from $x = 0$ m to $x = 10000$ m in the horizontal

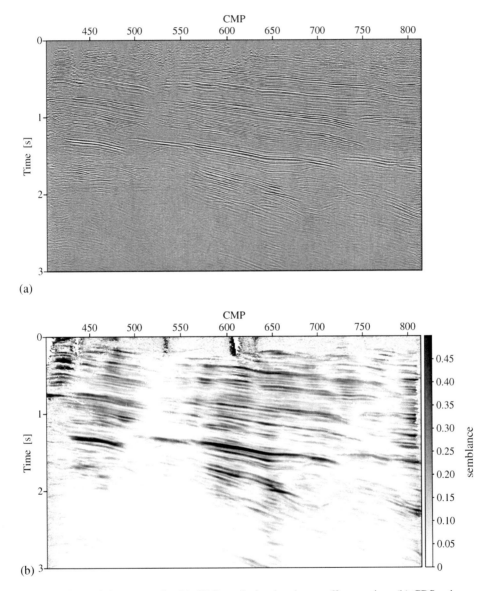

Figure 6.12: Real data example. (a) CRS stack simulated zero-offset section, (b) CRS coherence section.

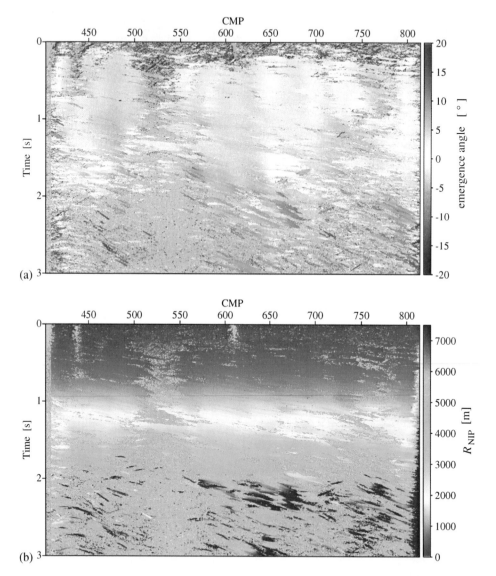

Figure 6.13: Real data example. (a) Emergence angle $\alpha$ section, (b) NIP wave radius $R_{\text{NIP}}$ section.

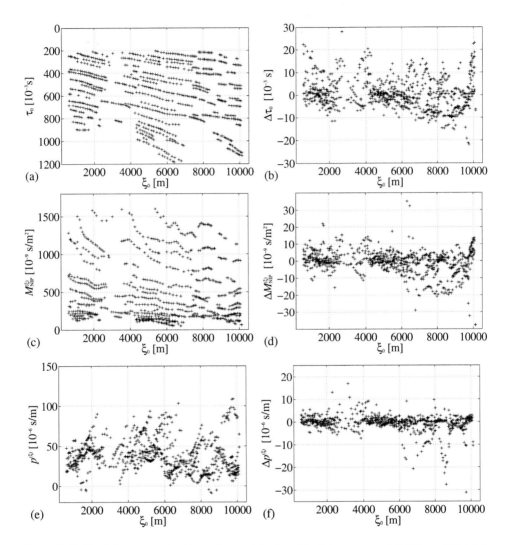

Figure 6.14: Real data example. Input data and final residual data errors. (a) Values of the data component $\tau_0$ used in the tomographic inversion. (b) Residual errors in $\tau_0$ after 10 iterations. (c) Values of the data component $M_{NIP}^{(\xi)}$ used in the tomographic inversion. (d) Residual errors in $M_{NIP}^{(\xi)}$ after 10 iterations. (e) Values of the data component $p^{(\xi)}$ used in the tomographic inversion. (f) Residual errors in $p^{(\xi)}$ after 10 iterations.

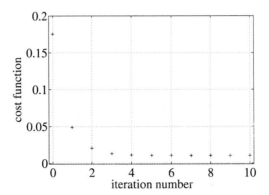

Figure 6.15: Real data example. Value of the cost function (4.16) as a function of iteration number.

direction and from the measurement surface $z = 0$ m to a depth of 3000 m. The initial model for the inversion is defined by a near-surface velocity of 1700 m/s and a constant vertical velocity gradient of 0.5 s$^{-1}$.

During the tomographic inversion, all of the different types of additional constraints discussed in Section 4.5 are applied. In order to avoid fluctuations of velocity in the shallowest part of the model where no data picks are placed, a near-surface velocity value of 1700 m/s is introduced as a priori information at a total of 23 model points with $z = 0$ and a constant horizontal spacing of 500 m. The corresponding data weight is set to $\sigma_v = 1$ (see Section 5.2). In addition, the constraint of minimum local velocity gradient along the reflector elements associated with the considered NIPs is used with a weight of $\sigma_{v_q} = 10$ (see Section 5.2).

To better constrain the velocity model at all of its boundaries, the spatially variable regularization described in Section 4.5 is applied. For that purpose, the coefficients $\varepsilon_{ij}^{xx}$ in equation (C.15) are increased by a factor 100 at the model boundaries ($i = 1, i = n_x, j = 1, j = n_z$) relative to the remaining coefficients. Away from the model boundaries, the regularization is the same for the horizontal and the vertical directions ($\varepsilon_{xx} = \varepsilon_{zz}$). As in all examples discussed above, the overall regularization weight is decreased from iteration to iteration according to equation (5.20). A suitable initial value for the overall regularization is found interactively by performing the inversion several times with a range of different initial regularization weights.

The averaging of higher spatial velocity derivatives around the normal ray during dynamic ray tracing and ray perturbation calculations (Section 5.2) is applied with a maximum averaging width of 500 m at $z = 0$.

With these parameters, a total of 10 nonlinear inversion iterations are performed. As can be seen in Figure 6.15, there are no significant changes in the cost function $S$ for iteration numbers larger than 6. The final residual misfit in the data components $\tau_0$, $M_{\mathrm{NIP}}^{(\xi)}$, and $p^{(\xi)}$ is displayed in Figures 6.14b, d, and f. Compared to the synthetic data example of Section 6.2, the residual data misfit remains relatively large. Obviously, the input data derived from the real seismic data are not as reliable as

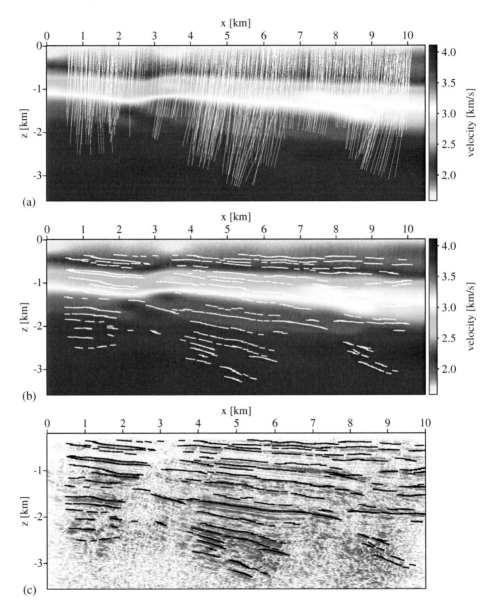

Figure 6.16: Real data example. Inversion result after 10 iterations. (a) Reconstructed smooth velocity model and normal rays corresponding to the final NIP model parameters. (b) Reconstructed
velocity model and local reflector elements representing the final NIP model parameters obtained
during the tomographic inversion. (c) Local reflector elements obtained during the tomographic
inversion, plotted into the prestack depth migration result obtained with the tomographic velocity
model.

those derived from the synthetic data. Possibly, not in all cases the same peak of the source signal has been picked and not all picked events are true primary reflections. More importantly, due to the large number of faults in the subsurface, the moveout of many of the picked reflection events is likely to deviate somewhat from hyperbolic, leading to inaccuracies in the data component $M_{\mathrm{NIP}}^{(\xi)}$.

The inversion result itself is displayed in Figures 6.16a and b. Figure 6.16a shows the obtained smooth velocity model together with the 793 normal rays corresponding to the input data. The associated final NIP model parameters are displayed in Figure 6.16b as local reflector elements representing the location and dip associated with each considered NIP. These reflector elements already give an idea of where reflection events will be positioned in the migrated image. Figure 6.16c shows the same local reflector elements plotted into the prestack depth migration result (see below) obtained with the velocity model of Figures 6.16a and b.

Obviously, certain parts of the subsurface, especially below 2 km depth are not covered by any of the normal rays associated with the picked data points. These areas correspond to the regions of low coherence in the CRS stack results discussed above. No reliable attributes can be expected in these regions and, therefore, no picks have been placed there. The corresponding parts of the velocity model (Figure 6.16a and b) can, thus, not be expected to be reliable. Velocities there are mainly controlled by the regularization constraints. In the central region of the model, on the other hand, sufficient ray coverage is available down to 3 km depth. That part of the model should therefore be well determined.

### 6.3.4 Depth migration

The smooth velocity model obtained with the tomographic inversion can now be used to perform a prestack depth migration. As with the synthetic data example of Section 6.2, a Kirchhoff migration algorithm (Hertweck, 2004) based on eikonal traveltimes is used. Figure 6.17 shows a stack of common-offset migrations for the offset range between 0 and 2000 m (see also Figure 6.16c). Particularly in the upper 2 km, but also in deeper regions, a number of steep faults have been well resolved. As expected, those parts of the subsurface below 2 km depth, that correspond to the regions of low coherence in the CRS stack section (Figure 6.12a), also show little coherent signal in the prestack depth migration result, while the central part of the section is well imaged up to 3 km depth. The deep events in the right part of the section, between $x = 8$ km and $x = 9$ km probably correspond to out-of-plane reflections.

A number of common-image gathers at selected image locations with a regular separation of 1 km along the section are shown in Figure 6.18. The displayed offsets range from 0 to 2000 m (which coincides with the offset range used during the CRS stack to determine the kinematic wavefield attributes). Most events in the CIGs are flat, indicating that the determined velocity model agrees well with the data. Some residual moveout is visible in the stack of shallow reflectors around 1 km depth. Also, the strong reflector at about 2 km depth exhibits a slight negative moveout in the right part of the section. The deep reflectors in the central part of the section are, however, perfectly flat.

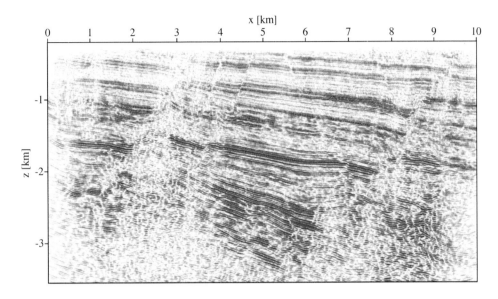

Figure 6.17: Real data example. Stack of common-offset Kirchhoff prestack depth migrations obtained using the smooth tomographic velocity model. The stacked offsets range from 0 to 2000 m. The applied offset-dependent mute is the same as in Figure 6.18.

All in all, the result is very encouraging. A suitable velocity model for depth migration has been determined from the seismic data without the need to apply repeated prestack migrations and with an extremely limited picking effort.

An automated picking procedure based on the CRS coherence section as outlined in Section 6.1 has also been applied to this dataset, yielding results very similar to those presented here. In general, however, completely automatic picking demands relatively good quality of the seismic data and the absence of multiples. Automatically picked data points always need to be carefully checked for outliers and multiples.

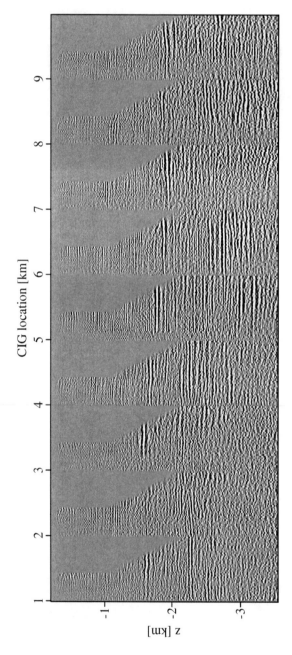

Figure 6.18: Real data example. Common-image gathers obtained by Kirchhoff prestack depth migration using the smooth tomographic velocity model. The maximum displayed offset is 2000 m.

# Chapter 7

# Conclusions and outlook

In this thesis, a new tomographic inversion method for the construction of smooth isotropic velocity models for seismic depth imaging has been presented. The method makes use of traveltime information in the form of kinematic wavefield attributes—coefficients of second-order traveltime approximations—extracted from the seismic multicoverage data with the common-reflection-surface (CRS) stack. The use of these attributes and the particular model parametrization in terms of isolated normal-incidence points (NIPs) in a smooth velocity model parameterized by B-splines leads to a number of distinct advantages compared to conventional tomographic and migration-based velocity estimation methods, especially in the case of seismic data with a low S/N ratio.

The weak point of conventional reflection tomography is the need for picking reflection events in the prestack data in order to obtain the required traveltime information for the tomographic inversion. This picking, which often needs to be performed along continuous interpreted horizons, is very time-consuming, especially in 3D seismic data, and can become difficult or even impossible when the overall S/N ratio in the data is low.

With the inversion approach based on kinematic wavefield attributes, event picking in the prestack data is avoided. Picking can be performed in the simulated zero-offset section/volume obtained with the CRS stack which has a significantly improved S/N ratio compared to the prestack data and is much better suited for picking. Information on the offset-dependence of traveltimes is contained in the kinematic wavefield attributes associated with each zero-offset sample. The attributes of the picked zero-offset samples are automatically extracted from the CRS stack results and can be directly used to perform the inversion.

The fact that each set of kinematic wavefield attributes used in the tomographic inversion can be associated with a single NIP in the subsurface and the fact that a smooth velocity model description is used, allow to treat pick locations in the data independently of each other. No continuous horizons need to be interpreted and it is sufficient to pick locally coherent events in the stacked section/volume. This is a clear advantage in situations of low S/N ratio, where it is often impossible to identify reflection events that are continuous and visible on all traces.

For the application of the tomographic inversion based on kinematic wavefield attributes, only a relatively small number of picks is required. This makes the entire inversion process, including the picking and the numerical calculations, very efficient. Under favorable conditions (good data quality) it is possible to automatize the picking, based on the CRS coherence section/volume, leading to a further gain in efficiency.

However, the use of second-order traveltime approximations also leads to limitations in the applicability of the attribute-based tomographic inversion in situations of strong lateral velocity variations. While it has been demonstrated on a number of synthetic and real data examples, including those shown in this thesis, that the method is in principle capable of handling lateral variations of velocity, it will fail if these variations become too severe and the kinematic wavefield attributes obtained with the CRS stack are no longer reliable. Such situations can be identified by examining the CRS coherence section/volume and selected CMP gathers.

In this thesis, the complete theory of the new tomographic inversion method for the 1D, 2D, and 3D case has been presented. Implementational and practical aspects of the method have been discussed and the entire inversion procedure has been demonstrated on a synthetic and a real 2D seismic dataset.

In the future, the method will need to be applied and further tested on real 3D seismic data with limited, as well as with full azimuth coverage. The advantages of the tomographic inversion based on kinematic wavefield attributes—the efficient extraction and use of traveltime information for velocity model building—further gain in relevance in the 3D case, due to the huge amounts of data normally involved.

An issue that requires further research is that of identifying and avoiding data points related to multiple reflections. Such data points, if not removed, can seriously degrade the results of the tomographic inversion. In simple situations, multiples can be directly identified on grounds of their attribute values. In more complex situations this is no longer possible and other ways of solving the problem need to be found. Practical aspects that need to be investigated to further increase the degree of automatization of the inversion method include an improved automatic picking procedure, suitable criteria for editing picked input data, and an automatic regularization strategy.

# Appendix A

# Physical interpretation of kinematic wavefield attributes

The coefficients in the traveltime approximations (3.7) and (3.11) used during the CRS stack have a physical interpretation in terms of kinematic properties of two hypothetical wavefronts emerging at the measurement surface location $\boldsymbol{\xi}_0$. These are the so-called normal wave and the normal-incidence-point (NIP) wave (e. g., Hubral, 1983). In this appendix the relation between normal and NIP wave properties on the one hand and the coefficients in equations (3.7) and (3.11) on the other hand will be derived.

## A.1  Second-order traveltime approximations

The second-order approximation of reflection traveltime $t$ with respect to midpoint and half-offset coordinates around $\boldsymbol{\xi}_m = \boldsymbol{\xi}_0$ and $\mathbf{h} = \mathbf{0}$ is given by

$$t(\boldsymbol{\xi}_0 + \Delta\boldsymbol{\xi}, \mathbf{h}) = t_0 + \frac{\partial t}{\partial \boldsymbol{\xi}} \Delta\boldsymbol{\xi} + \frac{1}{2}\Delta\boldsymbol{\xi}^T \frac{\partial^2 t}{\partial \boldsymbol{\xi}^2} \Delta\boldsymbol{\xi} + \frac{1}{2}\mathbf{h}^T \frac{\partial^2 t}{\partial \mathbf{h}^2} \mathbf{h}, \tag{A.1}$$

where $\Delta\boldsymbol{\xi} = \boldsymbol{\xi}_m - \boldsymbol{\xi}_0$ and the zero-offset traveltime measured at $\boldsymbol{\xi}_0$ is denoted by $t_0$. The partial derivatives with respect to the midpoint and half-offset vectors symbolize the partial derivatives with respect to the corresponding vector components, taken at $\boldsymbol{\xi}_m = \boldsymbol{\xi}_0$ and $\mathbf{h} = \mathbf{0}$. Thus, $\partial t/\partial \boldsymbol{\xi}$ is a two-component vector and $\partial^2 t/\partial \boldsymbol{\xi}^2$ and $\partial^2 t/\partial \mathbf{h}^2$ are symmetric $2 \times 2$ matrices. Note that (A.1) contains no linear term in the offset coordinate $\mathbf{h}$. This is due to the principle of reciprocity, that is, the invariance of reflection traveltimes with respect to interchanging sources and receivers (replacing $\mathbf{h}$ by $-\mathbf{h}$). Reflection traveltime must therefore be an even function of $\mathbf{h}$.

By squaring equation (A.1) and neglecting higher order terms, a second-order approximation of $t^2$ may be obtained:

$$t^2(\boldsymbol{\xi}_0 + \Delta\boldsymbol{\xi}, \mathbf{h}) = \left(t_0 + \frac{\partial t}{\partial \boldsymbol{\xi}} \Delta\boldsymbol{\xi}\right)^2 + t_0 \left(\Delta\boldsymbol{\xi}^T \frac{\partial^2 t}{\partial \boldsymbol{\xi}^2} \Delta\boldsymbol{\xi} + \mathbf{h}^T \frac{\partial^2 t}{\partial \mathbf{h}^2} \mathbf{h}\right). \tag{A.2}$$

In the following, the traveltime derivatives $\partial t / \partial \boldsymbol{\xi}$, $\partial^2 t / \partial \boldsymbol{\xi}^2$, and $\partial^2 t / \partial \mathbf{h}^2$ at $\boldsymbol{\xi}_0$ will be related to quantities describing an emerging hypothetical normal wavefront and an emerging hypothetical NIP wavefront at $\boldsymbol{\xi}_0$.

## A.2 Normal wave

In order to physically interpret the partial derivatives associated with the midpoint coordinates, expression (A.1) will first be restricted to $\mathbf{h} = \mathbf{0}$. The resulting zero-offset traveltime formula reads

$$t(\boldsymbol{\xi}_0 + \Delta \boldsymbol{\xi}) = t_0 + \frac{\partial t}{\partial \boldsymbol{\xi}} \Delta \boldsymbol{\xi} + \frac{1}{2} \Delta \boldsymbol{\xi}^T \frac{\partial^2 t}{\partial \boldsymbol{\xi}^2} \Delta \boldsymbol{\xi} \; . \tag{A.3}$$

It is usually assumed that reflection traveltimes in zero-offset sections correspond to reflected waves with coincident down- and upgoing raypaths. This assumption, which implies normal incidence at the reflection point, holds if the lateral velocity variation in the subsurface is not too strong. If both sides of equation (A.3) are divided by two, the resulting expression may be viewed as describing the second-order one-way traveltime $\tau = t/2$ of a wave emerging at $\boldsymbol{\xi}_0$ due to an exploding reflector element at the normal-incidence point (NIP) of the central zero-offset ray in the subsurface. As the rays associated with this wave are locally normal to the reflector element, it is also known as the *normal wave*, (e. g., Hubral, 1983).

Writing the first and second spatial traveltime derivatives of this one-way wave on the measurement plane as $\mathbf{p}^{(\xi)}$ and $\underline{\mathbf{M}}_N^{(\xi)}$ yields a traveltime expression identical to the paraxial traveltime approximation (2.93) in Section 2.7:

$$\tau(\boldsymbol{\xi}_0 + \Delta \boldsymbol{\xi}) = \tau_0 + \mathbf{p}^{(\xi)} \cdot \Delta \boldsymbol{\xi} + \frac{1}{2} \Delta \boldsymbol{\xi}^T \underline{\mathbf{M}}_N^{(\xi)} \Delta \boldsymbol{\xi} \; , \tag{A.4}$$

where $\tau_0$ is the one-way traveltime along the central normal ray (the normal ray that starts at the NIP) measured at $\boldsymbol{\xi}_0$. The vector $\mathbf{p}^{(\xi)}$ contains the horizontal components of the normal ray slowness vector at $\boldsymbol{\xi}_0$. If the near-surface velocity $v_0$ at the location $\boldsymbol{\xi}_0$ is known and locally constant, the emergence direction of that ray may be computed. Equation (2.94) in Section 2.7 then allows to relate $\underline{\mathbf{M}}_N^{(\xi)}$ to the matrix of second traveltime derivatives of the normal wave in ray-centered coordinates, which can, with equation (2.68), be written in terms of the matrix $\underline{\mathbf{K}}_N$ of wavefront curvature of the normal wave. Thus, if the azimuth and emergence angles of the normal ray at $\boldsymbol{\xi}_0$ are given by $\psi$ and $\alpha$ (see Figure A.1), the coefficients of equation (A.1) may be written in terms of kinematic properties of the normal wave emerging at $\boldsymbol{\xi}_0$:

$$\frac{\partial t}{\partial \boldsymbol{\xi}} = 2 \mathbf{p}^{(\xi)} = \frac{2}{v_0} (\sin \alpha \cos \psi, \sin \alpha \sin \psi)^T$$

$$\frac{\partial^2 t}{\partial \boldsymbol{\xi}^2} = 2 \underline{\mathbf{M}}_N^{(\xi)} = \frac{2}{v_0} \mathbf{H} \mathbf{K}_N \mathbf{H}^T \; . \tag{A.5}$$

The matrix $\underline{\mathbf{H}}$, which is the upper left $2 \times 2$ submatrix of the transformation matrix $\hat{\mathbf{H}}$ from ray-centered Cartesian coordinates to the global Cartesian coordinate system associated with the measurement surface (see Section 2.7), also depends on $\psi$ and $\alpha$. If the $q_1$-direction of the ray-centered

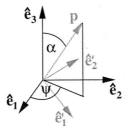

Figure A.1: Angles $\alpha$ and $\psi$ giving the emergence direction of the normal ray at $\boldsymbol{\xi}_0$. These angles may be used to define the orientation of the local ray-centered Cartesian coordinate system relative to the global Cartesian coordinate system on the measurement surface.

coordinate system, defined by the vector $\hat{\mathbf{e}}_1'$ in Figure A.1, is assumed to lie in the vertical plane spanned by the normal ray slowness vector and the measurement surface normal, matrix $\underline{\mathbf{H}}$ is given by

$$\underline{\mathbf{H}} = \begin{pmatrix} \cos\psi\cos\alpha & -\sin\psi \\ \sin\psi\cos\alpha & \cos\psi \end{pmatrix} . \tag{A.6}$$

## A.3   NIP wave

In order to find a physical interpretation of $\partial^2 t/\partial\mathbf{h}^2$, expression (A.1) is restricted to $\boldsymbol{\xi}_m = \boldsymbol{\xi}_0$ (a single CMP gather):

$$t(\mathbf{h}) = t_0 + \frac{1}{2}\mathbf{h}^T\frac{\partial^2 t}{\partial\mathbf{h}^2}\mathbf{h} . \tag{A.7}$$

It has been shown, among others, by Chernyak and Gritsenko (1979) and Hubral and Krey (1980), that up to second order in the offset coordinate, reflection traveltimes observed in a CMP gather co-incide with the traveltimes that would be observed if all involved rays passed through the normal-incidence point (NIP) of the zero-offset ray on the reflector (Figure A.2). The traveltime of a reflection event at an offset $\mathbf{h}$ in a CMP gather can therefore be calculated as the sum of the traveltimes along the two rays connecting the NIP with the surface locations $\boldsymbol{\xi}_0 + \mathbf{h}$ and $\boldsymbol{\xi}_0 - \mathbf{h}$. This statement is known as the *NIP wave theorem*.

The second-order traveltime at $\boldsymbol{\xi}_0 \pm \mathbf{h}$ due to a point source at the NIP can be expressed in terms of properties measured on the normal ray at $\boldsymbol{\xi}_0$ by equation (2.93):

$$\tau(\boldsymbol{\xi}_0 \pm \mathbf{h}) = \tau_0 \pm \mathbf{p}^{(\xi)}\cdot\mathbf{h} + \frac{1}{2}\mathbf{h}^T\underline{\mathbf{M}}_{\mathrm{NIP}}^{(\xi)}\mathbf{h} . \tag{A.8}$$

Here, $\underline{\mathbf{M}}_{\mathrm{NIP}}^{(\xi)}$ is the $2 \times 2$ matrix of second spatial derivatives of traveltime on the measurement plane due to a point source at the NIP. Again, with a known value for the near-surface velocity $v_0$, the matrix $\underline{\mathbf{M}}_{\mathrm{NIP}}^{(\xi)}$ can be related to the corresponding matrix of second spatial traveltime derivatives in

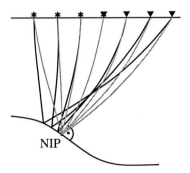

Figure A.2: The NIP wave theorem states that to second order in the offset coordinate, the CMP reflection traveltimes and the traveltimes along rays passing through the NIP of the zero-offset ray are identical.

ray-centered coordinates using (2.94), which is in turn linked to the matrix of wavefront curvature $\underline{\mathbf{K}}_{\text{NIP}}$ through equation (2.68). The wave due to a point source at the NIP is also known as the *NIP wave*.

As stated above, CMP traveltimes can be obtained as the sum of the traveltimes along the two ray segments connecting the NIP with the surface locations $\boldsymbol{\xi}_0 \pm \mathbf{h}$, given to second order by equation (A.8). Thus,

$$t(\mathbf{h}) = \tau(\boldsymbol{\xi}_0 - \mathbf{h}) + \tau(\boldsymbol{\xi}_0 + \mathbf{h}) = t_0 + \mathbf{h}^T \underline{\mathbf{M}}_{\text{NIP}}^{(\xi)} \mathbf{h} , \tag{A.9}$$

where $t_0 = 2\,\tau_0$ has been used. Comparing equations (A.9) and (A.7) yields

$$\frac{\partial^2 t}{\partial \mathbf{h}^2} = 2\underline{\mathbf{M}}_{\text{NIP}}^{(\xi)} = \frac{2}{v_0}\underline{\mathbf{H}}\mathbf{K}_{\text{NIP}}\underline{\mathbf{H}}^T , \tag{A.10}$$

where $\underline{\mathbf{H}}$ is the same matrix as in equation (A.5). Inserting expressions (A.5) and (A.10) in equation (A.2) yields the 3D CRS operator in terms of normal wave and NIP wave properties.

## A.4 Kinematic wavefield attributes in the 2D case

If the acquisition of seismic data is restricted to a single straight line (2D acquisition), it is no longer possible to determine all components of the vector $\mathbf{p}^{(\xi)}$ and the matrices $\underline{\mathbf{M}}_N^{(\xi)}$ and $\underline{\mathbf{M}}_{\text{NIP}}^{(\xi)}$ from the data. To simplify further calculations in this case, the Cartesian coordinate system on the measurement surface is rotated and shifted horizontally to make its first horizontal axis (corresponding to $\hat{\mathbf{e}}_1$ in Figure A.1) fall onto the seismic profile. In this new coordinate system, the vectors $\boldsymbol{\xi}_m$, $\boldsymbol{\xi}_0$, and $\mathbf{h}$ may then be written as

$$\boldsymbol{\xi}_m = \begin{pmatrix} \xi_m \\ 0 \end{pmatrix} , \quad \boldsymbol{\xi}_0 = \begin{pmatrix} \xi_0 \\ 0 \end{pmatrix} , \text{ and } \quad \mathbf{h} = \begin{pmatrix} h \\ 0 \end{pmatrix} . \tag{A.11}$$

Accordingly, the expressions

$$\Delta\boldsymbol{\xi}^T \underline{\mathbf{M}}_{\mathrm{N}}^{(\xi)} \Delta\boldsymbol{\xi} \quad \text{and} \quad \mathbf{h}^T \underline{\mathbf{M}}_{\mathrm{NIP}}^{(\xi)} \mathbf{h} \tag{A.12}$$

reduce to

$$\left(M_{N}^{(\xi)}\right)_{11} \Delta\xi^2 \quad \text{and} \quad \left(M_{NIP}^{(\xi)}\right)_{11} h^2 , \tag{A.13}$$

where $\left(M_{N}^{(\xi)}\right)_{11}$ and $\left(M_{NIP}^{(\xi)}\right)_{11}$ are the upper left components of the matrices $\underline{\mathbf{M}}_{\mathrm{N}}^{(\xi)}$ and $\underline{\mathbf{M}}_{\mathrm{NIP}}^{(\xi)}$, respectively. From equations (A.5), (A.10), and (A.6) it follows that if $v_0$ is known and locally constant, these matrix components can be written in terms of the components of the wavefront curvature matrices $\underline{\mathbf{K}}_{\mathrm{N}}$ and $\underline{\mathbf{K}}_{\mathrm{NIP}}$ as

$$\left(M_{N}^{(\xi)}\right)_{11} = \frac{1}{v_0}\left[\cos^2\psi\cos^2\alpha\left(K_{\mathrm{N}}\right)_{11} - 2\sin\psi\cos\psi\cos\alpha\left(K_{\mathrm{N}}\right)_{12} + \sin^2\psi\left(K_{\mathrm{N}}\right)_{22}\right] \tag{A.14}$$

and

$$\left(M_{NIP}^{(\xi)}\right)_{11} = \frac{1}{v_0}\left[\cos^2\psi\cos^2\alpha\left(K_{\mathrm{NIP}}\right)_{11} - 2\sin\psi\cos\psi\cos\alpha\left(K_{\mathrm{NIP}}\right)_{12} + \sin^2\psi\left(K_{\mathrm{NIP}}\right)_{22}\right] , \tag{A.15}$$

where $\psi$ now defines the azimuth of the normal ray emergence direction at $\boldsymbol{\xi}_0$ in the new rotated and shifted coordinate system. Defining $M_{N}^{(\xi)} := \left(M_{N}^{(\xi)}\right)_{11}$, $M_{NIP}^{(\xi)} := \left(M_{NIP}^{(\xi)}\right)_{11}$, and $p^{(\xi)} := p_1^{(\xi)}$ yields the CRS operator restricted to a single line:

$$t^2(\xi_0 + \Delta\xi, h) = \left(t_0 + 2p^{(\xi)}\Delta\xi\right)^2 + 2t_0\left(M^{(\xi)}\Delta\xi^2 + M_{NIP}^{(\xi)}h^2\right) . \tag{A.16}$$

If the azimuth of the normal ray emergence direction coincides with the direction of the seismic line ($\psi = 0$) only the first term on the right-hand sides of equations (A.14) and (A.15) remains. This is, for instance, the case if subsurface structures are invariant in the direction perpendicular to the acquisition line direction (2.5 D case, e. g., Bleistein, 1986). Defining $K_{\mathrm{N}} := \left(K_{\mathrm{N}}\right)_{11}$ and $K_{\mathrm{NIP}} := \left(K_{\mathrm{NIP}}\right)_{11}$ then results in the expressions

$$M_{N}^{(\xi)} = \frac{\cos^2\alpha}{v_0} K_{\mathrm{N}} ,$$

$$M_{NIP}^{(\xi)} = \frac{\cos^2\alpha}{v_0} K_{\mathrm{NIP}} , \tag{A.17}$$

$$p^{(\xi)} = \frac{\sin\alpha}{v_0} .$$

With the expressions (A.17), the CRS operator restricted to a single line for the special case of subsurface structures that are invariant in the direction perpendicular to that line reads

$$t^2(\xi_0 + \Delta\xi, h) = \left(t_0 + \frac{2\sin\alpha}{v_0}\Delta\xi\right)^2 + \frac{2t_0\cos^2\alpha}{v_0}\left(K_{\mathrm{N}}\Delta\xi^2 + K_{\mathrm{NIP}}h^2\right) . \tag{A.18}$$

# Appendix B

# Velocity model description with B-splines

The tomographic inversion method introduced in Chapter 4 makes use of a velocity model description based on B-splines (de Boor, 1978). In this appendix some properties of B-splines will be briefly summarized.

## B.1 Definition and properties

B-splines of degree $m$ are spline basis functions which allow any spline function $v(x)$ of the same degree, defined by its values at a sequencee of knot locations, to be represented in the form of a weighted sum

$$v(x) = \sum_i v_i \beta_i(x) \; . \tag{B.1}$$

Here, the $v_i$ are called B-spline coefficients and the $\beta_i(x)$ are the B-spline basis functions. On a given increasing knot sequence

$$\ldots < x_i < x_{i+1} < \ldots$$

(not necessarily with a constant interval length), B-splines are spline functions of minimum length. Being of degree $m$, they are non-zero only on $m+1$ consecutive knot intervals. B-splines can be defined recursively by (e. g., Oevel, 1996)

$$\beta^{[m]}_{[x_i,\ldots,x_{i+m+1}]}(x) = \frac{x - x_i}{x_{i+m} - x_i} \beta^{[m-1]}_{[x_i,\ldots,x_{i+m}]}(x) + \frac{x_{i+m+1} - x}{x_{i+m+1} - x_{i+1}} \beta^{[m-1]}_{[x_{i+1},\ldots,x_{i+m+1}]}(x) \tag{B.2}$$

with

$$\beta^{[0]}_{[x_i,x_{i+1}]}(x) = \begin{cases} 1 & \text{for } x \in [x_i, x_{i+1}) \\ 0 & \text{else} \; . \end{cases} \tag{B.3}$$

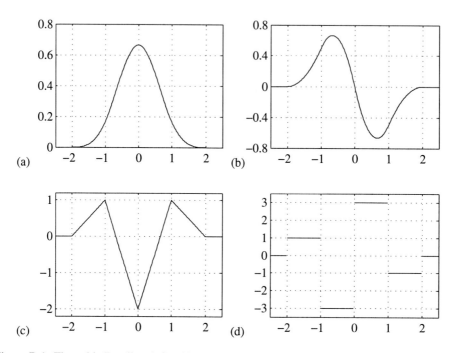

Figure B.1: The cubic B-spline defined in equation (B.4). (a) The B-spline function itself, (b) its first spatial derivative, (c) its second spatial derivative, and (d) its third spatial derivative.

With this definition, a B-spline of degree 3 (cubic B-spline) on the knot sequence $[-2,-1,0,1,2]$ is given by

$$
\beta^{[3]}_{[-2,-1,0,1,2]}(x) =
\begin{cases}
0 & x \le -2 \\
(2+x)^3/6 & x \in [-2,-1] \\
(4-6x^2-3x^3)/6 & x \in [-1,0] \\
(4-6x^2+3x^3)/6 & x \in [0,1] \\
(2-x)^3/6 & x \in [1,2] \\
0 & x \ge 2 .
\end{cases}
\tag{B.4}
$$

This cubic B-spline, together with its first, second and third derivatives is depicted in Figure B.1. It can be shown that the following properties of B-splines as defined in equations (B.2) and (B.3) hold for any degree $m \ge 0$ (Oevel, 1996):

a)    $\beta^{[m]}_{[x_i,\dots,x_{i+m+1}]}$    is a spline of degree $m$

b)    $\beta^{[m]}_{[x_i,\dots,x_{i+m+1}]}(x) = 0$    for $x \notin (x_i,\dots,x_{i+m+1})$,    $m \ge 1$

c) $\beta^{[m]}_{[x_i,\ldots,x_{i+m+1}]}(x) \in (0,1]$   for $x \in (x_i, x_{i+m+1})$

d) $\sum\limits_{i=-\infty}^{\infty} \beta^{[m]}_{[x_i,\ldots,x_{i+m+1}]}(x) = \sum\limits_{i=j-m}^{j} \beta^{[m]}_{[x_i,\ldots,x_{i+m+1}]}(x) = 1$   for $x \in [x_j, x_{i+m+1}]$

e) $\frac{d}{dx}\beta^{[m]}_{[x_i,\ldots,x_{i+m+1}]}(x) = m\left(\frac{1}{x_{i+m}-x_i}\beta^{[m-1]}_{[x_i,\ldots,x_{i+m}]}(x) - \frac{1}{x_{i+m+1}-x_{i+1}}\beta^{[m-1]}_{[x_{i+1},\ldots,x_{i+m+1}]}(x)\right),$   $m \geq 1$

An efficient recursive algorithm for the numerical evaluation of B-splines and its derivatives at a given location $x$ has been given by de Boor (1978).

## B.2   Velocity models in terms of B-splines

In the context of 2D and 3D tomographic inversion as described in Chapter 5, velocity models with continuous third derivatives are required (see Appendices D and E). Therefore, B-splines of degree $m = 4$ will be used to describe 2D and 3D velocity models, while B-splines of degree $m = 3$ will be used in the 1D case. When the degree $m$ has been fixed, a simplified notation for the B-spline functions will be used:

$$\beta_i(x) := \beta^{[m]}_{[x_{i-2},\ldots,x_{i+m-1}]} . \tag{B.5}$$

For the representation of a 1D function with $n_x$ B-spline coefficients, a total of $n_x + m + 1$ knots are required. Consequently, the function given in (B.1) with the coefficients $v_i$ defined at the knot locations $[x_1,\ldots,x_{n_x}]$ requires an additional $m+1$ knots, resulting in a sequence $[x_{-1},\ldots,x_{n_x+m-1}]$.

The $n$th derivative of a function which is defined by B-splines can be simply written as

$$\frac{d^n v(x)}{dx^n} = \sum_{i=1}^{n_x} v_i \frac{d^n \beta_i(x)}{dx^n} , \tag{B.6}$$

where the $n$th derivative of the B-spline functions is given by property e), above.

A B-spline representation of functions of higher dimensions is also possible. For the purposes of describing 2D velocity models, a grid of $(n_x + m - 1) \times (n_z + m - 1)$ knots is defined by two strictly increasing sequences, $[x_{-1},\ldots,x_{n_x+m-1}]$ and $[-z_{-1},\ldots,-z_{n_z+m-1}]$. Note that the positive $z$-direction points upwards, so that depth increases with increasing values of $-z$. Accordingly, the sequence $[-z_{-1},\ldots,-z_{n_z+m-1}]$ is strictly increasing with depth if $z_k > z_{k+1}$. A 2D velocity function can then be written in terms of B-splines as

$$v(x,z) = \sum_{i=1}^{n_x}\sum_{k=1}^{n_z} v_{ik}\beta_i(x)\beta_k(-z)$$

$$= \sum_{i=1}^{n_x} v_i(z)\beta_i(x) \quad \text{with} \quad v_i(z) = \sum_{k=1}^{n_z} v_{ik}\beta_k(-z) . \tag{B.7}$$

Analogously, functions in three or more dimensions can be defined. The generalization of the differentiation rule (B.6) to higher dimensions is straightforward. In the 2D case, derivatives of

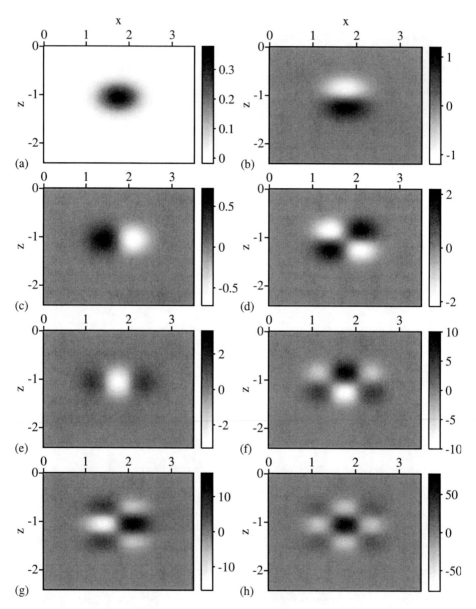

Figure B.2: A two-dimensional B-spline function of degree $m = 4$, as defined in equation (B.7), with only one non-zero coefficient $v_{ik}$. The horizontal knot spacing is 0.5 while the vertical knot spacing is 0.3. Figures (a) to (h) show the B-spline function $v(x,z)$ itself, together with its spatial derivatives up to second order in $x$ and $z$: (a) $v(x,z)$, (b) $\partial v(x,z)/\partial z$, (c) $\partial v(x,z)/\partial x$, (d) $\partial^2 v(x,z)/\partial x \partial z$, (e) $\partial^2 v(x,z)/\partial x^2$, (f) $\partial^3 v(x,z)/\partial x^2 \partial z$, (g) $\partial^3 v(x,z)/\partial x \partial z^2$, (h) $\partial^4 v(x,z)/\partial x^2 \partial z^2$.

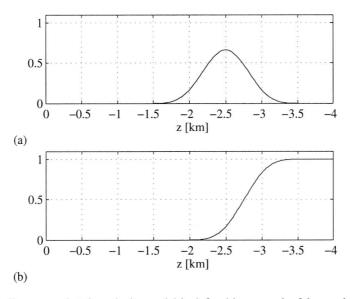

(a)

(b)

Figure B.3: To ensure, that the velocity model is defined in a meaningful way also towards the model boundaries, the B-spline knot interval lengths at the boundaries are set to very large values. Consequently, while B-spline basis functions in the central part of the model, (a), are localized around a certain spatial location, the outermost B-spline basis function, (b), remains virtually constant for large values of $-z$.

(B.7) can be written as

$$\frac{\partial^{(n+l)}v(x,z)}{\partial x^n \partial z^l} = \sum_{i=1}^{n_x} \sum_{k=1}^{n_z} v_{ik}(-1)^l \frac{\partial^n \beta_i(x)}{\partial x^n} \frac{\partial^l \beta_k(-z)}{\partial z^l} \ . \tag{B.8}$$

An example of a 2D function defined as in (B.7) with $m = 4$ and only one non-zero coefficient is depicted in Figure B.2 together with its derivatives up to second order in the vertical and horizontal directions. This function represents the basis element, from which velocity models are built in the 2D case. It also represents the velocity perturbation used in the application of 2D ray perturbation theory to calculate the Fréchet derivatives with respect to velocity for the tomographic matrix.

Velocity models described in terms of B-splines are non-zero only between the minimum and maximum used knot locations. In order to ensure that the velocity distributions obtained by tomographic inversion are defined in a meaningful way also outside of the region constrained by the data, the B-spline knot interval lengths at the margins of the model are set to very large values. The velocity, thus, remains virtually constant in a large region around the central part of the model (Figure B.3).

# Appendix C

# Regularization

The matrix involved in the calculation of a model update vector $\Delta\mathbf{m}$ from the data misfit vector $\Delta\mathbf{d}(\mathbf{m})$ as discussed in Section 4.4 is usually singular or near-singular, meaning that the corresponding equations do not contain sufficient information to uniquely determine all model parameters. To ensure a stable solution of the inverse problem, additional constraints on the model parameters need to be introduced. A physically sensible way of doing this is to require the velocity model to have minimum curvature, that is, minimum second derivatives, as a smooth model without artificial structure is sought. In order to make the solution of the inverse problem as much as possible independent of the B-spline knot interval spacing used in the velocity model description, the minimum curvature condition is imposed on the smooth velocity distribution itself, not on the model parameters (Delprat-Jannaud and Lailly, 1993). As a measure of the overall curvature of the velocity model, the integral of the square of its second derivatives over the entire spatial domain considered during the inversion is used.

## C.1  1D case

In the 1D case, the second spatial derivative of $v(z)$ written in terms of B-splines reads (see Appendix B)

$$\frac{\partial^2 v(z)}{\partial z^2} = \sum_{i=1}^{n_z} v_i \frac{\partial^2 \beta_i(-z)}{\partial z^2} \ . \tag{C.1}$$

If not only the second derivative of $v$, but also $v$ itself is considered for the regularization, the measure of model curvature to be used in the regularization term of the cost function can be written as

$$\int_z \left[ \varepsilon_{zz} \left( \frac{\partial^2 v(z)}{\partial z^2} \right)^2 + \varepsilon v^2(z) \right] dz = \sum_{i=1}^{n_z} \sum_{j=1}^{n_z} v_i v_j \int_z \left[ \varepsilon_{zz} \frac{\partial^2 \beta_i(-z)}{\partial z^2} \frac{\partial^2 \beta_j(-z)}{\partial z^2} + \varepsilon \beta_i(-z) \beta_j(-z) \right] dz$$

$$= \sum_{i=1}^{n_z} \sum_{j=1}^{n_z} v_i v_j D_{ij}''^{(1D)}$$

$$= \mathbf{m}^{(v)T} \underline{\mathbf{D}}''^{(1D)} \mathbf{m}^{(v)} .$$

(C.2)

Here, $\underline{\mathbf{D}}''^{(1D)} = \varepsilon_{zz} \hat{\underline{\mathbf{D}}}^{zz} + \varepsilon \hat{\underline{\mathbf{D}}}$ with elements

$$\hat{D}_{ij}^{zz} = \int_z \frac{\partial^2 \beta_i(-z)}{\partial z^2} \frac{\partial^2 \beta_j(-z)}{\partial z^2} dz \quad \text{and}$$

$$\hat{D}_{ij} = \int_z \beta_i(-z) \beta_j(-z) dz ,$$

(C.3)

where the integration is carried out over the entire depth range considered during the inversion and $\mathbf{m}^{(v)}$ is defined by $m_i^{(v)} := v_i$. The factors $\varepsilon_{zz} > 0$ and $\varepsilon \geq 0$ are used for normalization and relative weighting of the two terms in (C.2). The second term, weighted by $\varepsilon$, is required for further calculations (Section 4.4), therefore, $\varepsilon > 0$ will be assumed here. Expression (C.2) may then be considered a norm squared for $v(z)$. Otherwise (for $\varepsilon = 0$ ), matrix $\underline{\mathbf{D}}''$ is not positive definite and (C.2) represents a seminorm. The factor $\varepsilon$ should, however, be chosen much smaller than $\varepsilon_{zz}$, as the objective is to minimize the curvature of velocity and not velocity itself.

## C.2   2D case

In the 2D case, a two-dimensional integral over the weighted sum of squared velocity and the squared second derivatives of velocity will be used as a measure of velocity model curvature. This integral can be written as a sum of three integrals (mixed derivatives are not considered), which may again be expressed in terms of the velocity model parameters. The integral over the squared second derivative of velocity with respect to the $x$-coordinate reads

$$\int_z \int_x \left( \frac{\partial^2 v}{\partial x^2} \right)^2 dx \, dz = \int_z \int_x \left[ \sum_{i=1}^{n_x} \sum_{j=1}^{n_z} v_{ij} \frac{\partial^2 \beta_i(x)}{\partial x^2} \beta_j(-z) \sum_{k=1}^{n_x} \sum_{l=1}^{n_z} v_{kl} \frac{\partial^2 \beta_k(x)}{\partial x^2} \beta_l(-z) \right] dx \, dz$$

$$= \sum_{i=1}^{n_x} \sum_{j=1}^{n_z} \sum_{k=1}^{n_x} \sum_{l=1}^{n_z} v_{ij} v_{kl} \int_x \frac{\partial^2 \beta_i(x)}{\partial x^2} \frac{\partial^2 \beta_k(x)}{\partial x^2} dx \int_z \beta_j(-z) \beta_l(-z) \, dz$$

(C.4)

$$= \sum_{i=1}^{n_x} \sum_{j=1}^{n_z} \sum_{k=1}^{n_x} \sum_{l=1}^{n_z} v_{ij} v_{kl} \tilde{D}_{ik}^{xx} \hat{D}_{jl}$$

with

$$\tilde{D}_{ik}^{xx} = \int_x \frac{\partial^2 \beta_i(x)}{\partial x^2} \frac{\partial^2 \beta_k(x)}{\partial x^2} dx$$

(C.5)

and $\hat{D}_{jl}$ as defined in equation (C.3). If the B-spline coefficients $v_{ij}$ are arranged into a column vector $\mathbf{m}^{(v)}$ and matrices $\underline{\tilde{\mathbf{D}}}^{xx}$ and $\hat{\mathbf{D}}$ are combined into a $(n_x n_z) \times (n_x n_z)$ matrix $\underline{\mathbf{D}}^{xx}$ by defining

$$
\begin{aligned}
m^{(v)}_{[(i-1)n_z+j]} &:= v_{ij} \quad \text{and} \\
D^{xx}_{[(i-1)n_z+j],[(k-1)n_z+l]} &:= \tilde{D}^{xx}_{ik}\hat{D}_{jl} \,,
\end{aligned}
\tag{C.6}
$$

expression (C.4) may be written in a simplified way:

$$
\int_z \int_x \left(\frac{\partial^2 v}{\partial x^2}\right)^2 dx\,dz = \sum_{p=1}^{(n_x n_z)} \sum_{q=1}^{(n_x n_z)} m^{(v)}_p m^{(v)}_q D^{xx}_{pq} = \mathbf{m}^{(v)\,T}\underline{\mathbf{D}}^{xx}\mathbf{m}^{(v)} \,.
\tag{C.7}
$$

By defining additional matrices $\underline{\tilde{\mathbf{D}}}$ and $\hat{\underline{\mathbf{D}}}^{zz}$ with

$$
\begin{aligned}
\tilde{D}_{ik} &= \int_x \beta_i(x)\beta_k(x)\,dx \quad \text{and} \\
\hat{D}^{zz}_{jl} &= \int_z \frac{\partial^2 \beta_j(-z)}{\partial z^2}\frac{\partial^2 \beta_l(-z)}{\partial z^2}\,dz
\end{aligned}
\tag{C.8}
$$

and combining them with matrix $\hat{\mathbf{D}}$ to yield the $(n_x n_z) \times (n_x n_z)$ matrices $\underline{\mathbf{D}}^{zz}$ and $\underline{\mathbf{D}}$ with

$$
\begin{aligned}
D^{zz}_{[(i-1)n_z+j],[(k-1)n_z+l]} &= \tilde{D}_{ik}\hat{D}^{zz}_{jl} \quad \text{and} \\
D_{[(i-1)n_z+j],[(k-1)n_z+l]} &= \tilde{D}_{ik}\hat{D}_{jl} \,,
\end{aligned}
\tag{C.9}
$$

the overall measure of curvature of the 2D velocity model may be written in terms of the velocity model vector $\mathbf{m}^{(v)}$:

$$
\int_z \int_x \left[\varepsilon_{xx}\left(\frac{\partial^2 v(x,z)}{\partial x^2}\right)^2 + \varepsilon_{zz}\left(\frac{\partial^2 v(x,z)}{\partial z^2}\right)^2 + \varepsilon\, v^2(x,z)\right] dx\,dz = \mathbf{m}^{(v)\,T}\underline{\mathbf{D}}''^{(2D)}\mathbf{m}^{(v)} \,.
\tag{C.10}
$$

Here, $\underline{\mathbf{D}}''^{(2D)} = \varepsilon_{xx}\underline{\mathbf{D}}^{xx} + \varepsilon_{zz}\underline{\mathbf{D}}^{zz} + \varepsilon\underline{\mathbf{D}}$, where, again, $\varepsilon_{xx} > 0$, $\varepsilon_{zz} > 0$, and $\varepsilon \geq 0$ are normalization and weighting factors. As in the 1D case, $\varepsilon$ should be chosen much smaller than $\varepsilon_{xx}$ and $\varepsilon_{zz}$, but non-zero.

## C.3 3D case

In a way completely analogous to the 2D case, the measure of curvature for a 3D velocity model can be defined by

$$
\begin{aligned}
&\int_z \int_y \int_x \left[\varepsilon_{xx}\left(\frac{\partial^2 v(x,y,z)}{\partial x^2}\right)^2 + \varepsilon_{yy}\left(\frac{\partial^2 v(x,y,z)}{\partial y^2}\right)^2 + \varepsilon_{zz}\left(\frac{\partial^2 v(x,y,z)}{\partial z^2}\right)^2 + \varepsilon\, v^2(x,y,z)\right] dx\,dy\,dz \\
&= \mathbf{m}^{(v)\,T}\left(\varepsilon_{xx}\underline{\mathbf{D}}^{xx} + \varepsilon_{yy}\underline{\mathbf{D}}^{yy} + \varepsilon_{zz}\underline{\mathbf{D}}^{zz} + \varepsilon\underline{\mathbf{D}}\right)\mathbf{m}^{(v)} \\
&= \mathbf{m}^{(v)\,T}\underline{\mathbf{D}}''^{(3D)}\mathbf{m}^{(v)} \,.
\end{aligned}
\tag{C.11}
$$

Here, $\mathbf{m}^{(v)}$ is defined by

$$m^{(v)}_{[(i-1)n_y n_z + (j-1)n_z + k]} := v_{ijk} \tag{C.12}$$

and the $(n_x n_y n_z) \times (n_x n_y n_z)$ matrices $\underline{\mathbf{D}}^{xx}$, $\underline{\mathbf{D}}^{yy}$, $\underline{\mathbf{D}}^{zz}$, and $\underline{\mathbf{D}}$ in (C.11) are defined analogously to the corresponding $(n_x n_z) \times (n_x n_z)$ matrices in the 2D case. For example, matrix $\underline{\mathbf{D}}^{xx}$ is given by

$$D^{xx}_{[(i-1)n_y n_z + (j-1)n_z + k],[(l-1)n_y n_z + (m-1)n_z + n]} =$$
$$\int_x \frac{\partial^2 \beta_i(x)}{\partial x^2} \frac{\partial^2 \beta_l(x)}{\partial x^2} dx \int_y \beta_j(y) \beta_m(y) dy \int_z \beta_k(-z) \beta_n(-z) dz . \tag{C.13}$$

The matrices $\underline{\mathbf{D}}''^{(1D)}$, $\underline{\mathbf{D}}''^{(2D)}$, and $\underline{\mathbf{D}}''^{(3D)}$ are symmetric and positive definite if $\varepsilon > 0$, therefore each of them can be written as

$$\varepsilon'' \underline{\mathbf{D}}'' = \underline{\mathbf{B}}^T \underline{\mathbf{B}} , \tag{C.14}$$

where matrix $\underline{\mathbf{B}}$ is an upper triangular matrix with the same dimensions as $\underline{\mathbf{D}}''$ and $\varepsilon'' > 0$. This property of matrix $\underline{\mathbf{D}}''$ is required in Chapter 4. The factorization (C.14) of matrix $\underline{\mathbf{D}}''$ into an upper triangular matrix $\underline{\mathbf{B}}$ and its transpose $\underline{\mathbf{B}}^T$ can be obtained in a numerically very efficient way by *Cholesky decomposition* (e. g., Press et al., 1992).

## C.4   Spatially variable regularization

The factors $\varepsilon_{xx}$, $\varepsilon_{yy}$, $\varepsilon_{zz}$, and $\varepsilon$ may themselves be defined as spatially variable functions with a dependence on one or more spatial coordinates. This can be useful, if it is known beforehand, that parts of the velocity model will be underconstrained by the data, while in other parts there is sufficient information for obtaining a detailed velocity model. The overall regularization weight $\varepsilon''$ in the cost function (4.16) in Section 4.4 may then have a low value, while the spatially variable weights $\varepsilon_{xx}$, $\varepsilon_{yy}$, $\varepsilon_{zz}$ ensure sufficient constraints for all parts of the model.

An obvious way of describing the spatial dependence of the weight factors is, again, to use B-splines, defined on the same grid of knot locations as is used for the description of the velocity model itself. A spatially variable 2D weight function $\varepsilon_{xx}(x,z)$ may, for example, be defined as

$$\varepsilon_{xx}(x,z) = \sum_{i=1}^{n_x} \sum_{j=1}^{n_z} \varepsilon_{ij}^{xx} \beta_i(x) \beta_j(-z) , \tag{C.15}$$

where the coefficients $\varepsilon_{ij}^{xx}$ may be chosen based on a priori knowledge of where the model is well constrained by the data. Due to property d) of B-spline functions, given in Appendix B, choosing $\varepsilon_{ij}^{xx} = 1$ for all values of $i$ and $j$ is equivalent to using a constant weight factor $\varepsilon_{xx} = 1$.

Introducing a spatially variable weight function $\varepsilon_{xx}(x,z)$ as in equation (C.15) obviously leads to a redefinition of matrix $\underline{\mathbf{D}}''^{(2D)}$.

# Appendix D

# Fréchet derivatives for 2D tomographic inversion

For the iterative solution of the tomographic inverse problem discussed in Section 4.4, the derivatives of the modeling operator with respect to the model parameters are required. These quantities, also known as Fréchet derivatives (e. g., Tarantola, 1987), can be calculated during ray tracing along each considered ray by applying ray perturbation theory as introduced in Section 2.4.

In this appendix, the Fréchet derivatives for the 2D tomographic inversion, that is, the derivatives of the modeled quantities $(\tau_0, M_{\mathrm{NIP}}^{(\xi)}, p^{(\xi)}, \xi_0)_i$ with respect to the NIP model parameters $(x, z, \theta)_i^{(\mathrm{NIP})}$ and the velocity model parameters $v_{jk}$, defined in Section 5.2, are derived. As the forward modeling in the 2D case is performed with dynamic ray tracing in ray-centered coordinates (Section 2.5), ray perturbation theory will also be applied in these coordinates. The Fréchet derivatives for 1D tomographic inversion (Section 5.1) are then obtained as a special case of the 2D results.

For clarity of presentation, a simplified notation for the model and data components is used in this appendix: all subscripts and superscripts—except the superscript $(\xi)$—of these quantities are dropped. Consequently, the data components are denoted by $(\tau, M^{(\xi)}, p^{(\xi)}, \xi)$, while the NIP model components are given by $(x, z, \theta)$.

In accordance with Section 2.5, $q$ and $s$ denote the 2D ray-centered coordinates with $p$ being the slowness component associated with the $q$ direction. Quantities at the ray starting point in the subsurface (the NIP location) are written with a subscript 0, while the corresponding quantities at the ray end point (at the measurement surface) have a subscript 1. In order to calculate the Fréchet derivatives with respect to $x$, $z$, and $\theta$ from the results of ray perturbation theory, which is applied in ray-centered coordinates, it is necessary to transform perturbations of these quantities at the respective ray starting and end points into the corresponding perturbations in ray-centered coordinates.

If $\alpha$ denotes the ray emergence angle at the measurement surface and $z$ is defined positive upwards, the following relations can be derived from geometrical considerations (Figures D.1 and D.2):

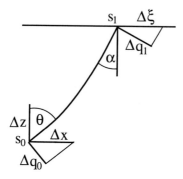

Figure D.1: Geometrical relation between perturbations of ray starting and end points in Cartesian coordinates to the corresponding quantities in ray-centered coordinates.

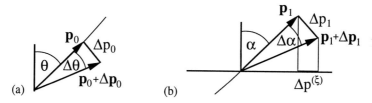

Figure D.2: Geometrical relation between perturbations of (a) ray starting angle and initial slowness component in ray-centered coordinates and (b) ray emergence angle and horizontal slowness component at the measurement surface.

$$\Delta\xi = \frac{1}{\cos\alpha}\Delta q_1 \,, \tag{D.1}$$

$$\Delta\alpha \approx \sin(\Delta\alpha) = v(s_1)\Delta p_1 \,, \tag{D.2}$$

$$\Delta p^{(\xi)} = \frac{\cos\alpha}{v(s_1)}\Delta\alpha = \cos\alpha\,\Delta p_1 \,, \tag{D.3}$$

$$\Delta q_0 = \cos\theta\,\Delta x - \sin\theta\,\Delta z \,, \tag{D.4}$$

$$\Delta p_0 = \frac{\sin(\Delta\theta)}{v(s_0)} \approx \frac{1}{v(s_0)}\Delta\theta \,, \tag{D.5}$$

$$\Delta s_0 = \sin\theta\,\Delta x + \cos\theta\,\Delta z \,. \tag{D.6}$$

# D.1 Fréchet derivatives of $\xi$ and $p^{(\xi)}$

Perturbations $\Delta q_1$ and $\Delta p_1$ of the ray-centered coordinate $q$ and associated slowness component $p$ at the ray end point due to perturbations of the corresponding quantities $\Delta q_0$ and $\Delta p_0$ at the ray starting point and perturbations $\Delta v$ of the velocity along the ray can be calculated with equation (2.45). Making use of equation (2.46), equation (2.45) can be written for 2D ray-centered coordinates as

$$\begin{pmatrix} \Delta q_1 \\ \Delta p_1 \end{pmatrix} = \mathbf{\Pi}(s_1, s_0) \begin{pmatrix} \Delta q_0 \\ \Delta p_0 \end{pmatrix} + \mathbf{\Pi}(s_1, s_0) \int_{s_0}^{s_1} \mathbf{\Pi}^{-1}(s, s_0) \Delta\mathbf{w}(s, \Delta v(s)) \, ds . \tag{D.7}$$

With the reduced Hamiltonian defined in equation (2.70) and with equation (2.42), the vector $\Delta\mathbf{w}$, equation (2.44), evaluated on the central ray can be written as

$$\Delta\mathbf{w} = \left( \frac{\partial \Delta H}{\partial p}, -\frac{\partial \Delta H}{\partial q} \right)^T = \left( 0, -\frac{1}{v^2}\frac{\partial \Delta v}{\partial q} + \frac{1}{v^3}\frac{\partial v}{\partial q}\Delta v \right)^T . \tag{D.8}$$

Using this expression in equation (D.7) yields

$$\begin{pmatrix} \Delta q_1 \\ \Delta p_1 \end{pmatrix} = \mathbf{\Pi}(s_1, s_0) \left[ \begin{pmatrix} \Delta q_0 \\ \Delta p_0 \end{pmatrix} + \begin{pmatrix} \Delta\tilde{q}(\Delta v) \\ \Delta\tilde{p}(\Delta v) \end{pmatrix} \right] \tag{D.9}$$

with

$$\begin{aligned}
\Delta\tilde{q}(\Delta v) &= -\int_{s_0}^{s_1} Q_2(s, s_0)\left( -\frac{1}{v^2}\frac{\partial \Delta v}{\partial q} + \frac{1}{v^3}\frac{\partial v}{\partial q}\Delta v \right) ds , \\
\Delta\tilde{p}(\Delta v) &= \int_{s_0}^{s_1} Q_1(s, s_0)\left( -\frac{1}{v^2}\frac{\partial \Delta v}{\partial q} + \frac{1}{v^3}\frac{\partial v}{\partial q}\Delta v \right) ds .
\end{aligned} \tag{D.10}$$

Substituting $\Delta q_1$ and $\Delta p_1$ from equation (D.9) into the expressions (D.1) and (D.3) and using equations (D.4) and (D.5) results in

$$\begin{aligned}
\Delta\xi = \ &\frac{\cos\theta}{\cos\alpha} Q_1(s_1, s_0)\Delta x \\
&- \frac{\sin\theta}{\cos\alpha} Q_1(s_1, s_0)\Delta z \\
&+ \frac{1}{v(s_0)\cos\alpha} Q_2(s_1, s_0)\Delta\theta \\
&+ \frac{1}{\cos\alpha}\left[ Q_1(s_1, s_0)\Delta\tilde{q}(\Delta v) + Q_2(s_1, s_0)\Delta\tilde{p}(\Delta v) \right]
\end{aligned} \tag{D.11}$$

and

$$\begin{aligned}
\Delta p^{(\xi)} = \ &\cos\alpha \cos\theta \ P_1(s_1, s_0)\Delta x \\
&- \cos\alpha \sin\theta \ P_1(s_1, s_0)\Delta z \\
&+ \frac{\cos\alpha}{v(s_0)} P_2(s_1, s_0)\Delta\theta \\
&+ \cos\alpha\left[ P_1(s_1, s_0)\Delta\tilde{q}(\Delta v) + P_2(s_1, s_0)\Delta\tilde{p}(\Delta v) \right] .
\end{aligned} \tag{D.12}$$

To obtain the derivatives of $\xi$ and $p^{(\xi)}$ with respect to the velocity model parameters $v_{jk}$ from these expressions, consider a linear functional $F$ such that $\Delta f = F(\Delta v)$. Taking into account the definition of the velocity model in terms of B-splines, equation (5.22), with the $v_{jk}$ acting as coefficients, it follows that the derivative of $f$ with respect to the $v_{jk}$ can be written as

$$\frac{\partial f}{\partial v_{jk}} = F\left(\frac{\partial v}{\partial v_{jk}}\right) = F\left(\beta_j(x)\beta_k(-z)\right) . \tag{D.13}$$

With this result applied in equations (D.11) and (D.12), the Fréchet derivatives of $\xi$ and $p^{(\xi)}$ with respect to the NIP and velocity model parameters can be written as follows:

$$\frac{\partial \xi_i}{\partial x_j} = \frac{\cos\theta_j}{\cos\alpha_j} Q_1(s_{1j}, s_{0j})\, \delta_{ij} ,$$

$$\frac{\partial \xi_i}{\partial z_j} = -\frac{\sin\theta_j}{\cos\alpha_j} Q_1(s_{1j}, s_{0j})\, \delta_{ij} ,$$

$$\frac{\partial \xi_i}{\partial \theta_j} = \frac{1}{v(s_{0j})\cos\alpha_j} Q_2(s_{1j}, s_{0j})\, \delta_{ij} ,$$

$$\frac{\partial \xi_i}{\partial v_{jk}} = \frac{1}{\cos\alpha_j}\left[Q_1(s_{1j}, s_{0j})\frac{\partial \tilde{q}_i}{\partial v_{jk}} + Q_2(s_{1j}, s_{0j})\frac{\partial \tilde{p}_i}{\partial v_{jk}}\right] ,$$

$$\frac{\partial p_i^{(\xi)}}{\partial x_j} = \cos\alpha_j\cos\theta_j\, P_1(s_{1j}, s_{0j})\, \delta_{ij} , \tag{D.14}$$

$$\frac{\partial p_i^{(\xi)}}{\partial z_j} = -\cos\alpha_j\sin\theta_j\, P_1(s_{1j}, s_{0j})\, \delta_{ij} ,$$

$$\frac{\partial p_i^{(\xi)}}{\partial \theta_j} = \frac{\cos\alpha_j}{v(s_{0j})} P_2(s_{1j}, s_{0j})\, \delta_{ij} ,$$

$$\frac{\partial p_i^{(\xi)}}{\partial v_{jk}} = \cos\alpha_j\left[P_1(s_{1j}, s_{0j})\frac{\partial \tilde{q}_i}{\partial v_{jk}} + P_2(s_{1j}, s_{0j})\frac{\partial \tilde{p}_i}{\partial v_{jk}}\right] ,$$

where

$$\frac{\partial \tilde{q}_i}{\partial v_{jk}} = -\int_{s_{0i}}^{s_{1i}} Q_2(s, s_{0i})\left\{-\frac{1}{v^2}\frac{\partial}{\partial q}\left[\beta_j(x(s))\beta_k(-z(s))\right] + \frac{1}{v^3}\frac{\partial v}{\partial q}\left[\beta_j(x(s))\beta_k(-z(s))\right]\right\} ds \tag{D.15}$$

and

$$\frac{\partial \tilde{p}_i}{\partial v_{jk}} = \int_{s_{0i}}^{s_{1i}} Q_1(s, s_{0i})\left\{-\frac{1}{v^2}\frac{\partial}{\partial q}\left[\beta_j(x(s))\beta_k(-z(s))\right] + \frac{1}{v^3}\frac{\partial v}{\partial q}\left[\beta_j(x(s))\beta_k(-z(s))\right]\right\} ds . \tag{D.16}$$

Here, an index $i$ denotes properties associated with the $i$th ray and $\delta_{ij}$ is the Kronecker symbol. It accounts for the fact that perturbations of quantities associated with a given ray have no influence on quantities of any other ray.

## D.2   Fréchet derivatives of $M^{(\xi)}$

The quantity $M^{(\xi)}$ is the second derivative of NIP wave traveltime with respect to the horizontal coordinate at the emergence location of the normal ray. If the near-surface velocity at that location can be assumed to be locally constant, $M^{(\xi)}$ can be related to the corresponding second traveltime derivative $M$ with respect to the coordinate $q$ by

$$M^{(\xi)} = M \cos^2 \alpha .\qquad (D.17)$$

Consequently, a perturbation of $M^{(\xi)}$ can be expressed in terms of perturbations of $\alpha$ and $M$:

$$\Delta M^{(\xi)} = -2 \sin \alpha \cos \alpha \, M \Delta \alpha + \cos^2 \alpha \, \Delta M .\qquad (D.18)$$

The quantity $M$, being related to a point source at the normal ray initial point $s_0$ (the NIP), can be written in terms of the elements $Q_2$ and $P_2$ of the ray propagator matrix $\underline{\underline{\Pi}}(s_1, s_0)$ associated with normalized point-source initial conditions: $M = P_2/Q_2$, equation (2.75). Perturbations of $M$ can, thus, be related to perturbations of the elements of matrix $\underline{\underline{\Pi}}(s_1, s_0)$:

$$\Delta M = \frac{1}{Q_2} \Delta P_2 - \frac{P_2}{Q_2^2} \Delta Q_2 .\qquad (D.19)$$

Perturbations of the ray propagator matrix can, in turn, be related to perturbations of $q$ and $p$ at the initial point $s_0$ of the ray and to velocity perturbations along the ray with the help of the ray perturbation theory expression (2.54) given in Section 2.4:

$$\Delta \underline{\underline{\Pi}} = \underline{\underline{\Pi}}(s_1, s_0) \int_{s_0}^{s_1} \underline{\underline{\Pi}}^{-1}(s, s_0) \Delta \underline{\underline{S}}(s) \underline{\underline{\Pi}}(s, s_0) \, ds ,\qquad (D.20)$$

where

$$\Delta \underline{\underline{S}}(s) = \Delta \underline{\underline{S}}_1 \big( \Delta v(s) \big) + \Delta \underline{\underline{S}}_2 \big( \Delta q(s), \Delta p(s) \big) .\qquad (D.21)$$

The matrices $\Delta \underline{\underline{S}}_1$ and $\Delta \underline{\underline{S}}_2$ can be calculated from equations (2.49) and (2.50) using the definition of the reduced Hamiltonian $H$ for 2D ray-centered coordinates, equation (2.70), and the definition of $\Delta H$, equation (2.42):

$$\Delta \underline{\underline{S}}_1 (\Delta v) = \begin{pmatrix} 0 & \Delta v \\ \frac{2}{v^2} \left( \frac{\Delta v}{v} \frac{\partial^2 v}{\partial q^2} + \frac{1}{v} \frac{\partial v}{\partial q} \frac{\partial \Delta v}{\partial q} - \frac{\Delta v}{v^2} \left( \frac{\partial v}{\partial q} \right)^2 - \frac{1}{2} \frac{\partial^2 \Delta v}{\partial q^2} \right) & 0 \end{pmatrix}\qquad (D.22)$$

and

$$\Delta \underline{\underline{S}}_2 (\Delta q, \Delta p) = \begin{pmatrix} 2 \frac{\partial v}{\partial q} \Delta p & 2 \frac{\partial v}{\partial q} \Delta q \\ \left( \frac{3}{v^3} \frac{\partial v}{\partial q} \frac{\partial^2 v}{\partial q^2} - \frac{1}{v^2} \frac{\partial^3 v}{\partial q^3} \right) \Delta q & -2 \frac{\partial v}{\partial q} \Delta p \end{pmatrix} .\qquad (D.23)$$

The quantities $\Delta q$ and $\Delta p$ in equation (D.23) are evaluated at the current location on the ray. They can be related to perturbations $\Delta q_0$ and $\Delta p_0$ at the ray starting point and perturbations $\Delta v$ of the velocity along the ray by equation (2.45). In practice, however, the dependence of $\Delta \underline{\underline{S}}_2$ on perturbations of velocity will be neglected. This means, that $\Delta q$ and $\Delta p$ in equation (D.23) are simply calculated from $\Delta q_0$ and $\Delta p_0$ using the ray propagator matrix (2.73).

It can easily be shown—using the symplectic properties of $\mathbf{\Pi}$, equations (2.30)—that, if perturbations of the ray propagator matrix can be written in the form

$$\Delta\mathbf{\Pi} = \mathbf{\Pi}\mathbf{A} = \begin{pmatrix} Q_1 & Q_2 \\ P_1 & P_2 \end{pmatrix}\begin{pmatrix} A_{11} & A_{12} \\ A_{21} & A_{22} \end{pmatrix}, \tag{D.24}$$

expression (D.19) for $\Delta M$ reduces to

$$\Delta M = -Q_2^{-2}A_{12}. \tag{D.25}$$

This means that for the calculation of $\Delta M$ only the upper right element of the matrix under the integral of equation (D.20) is required:

$$\Delta M(\Delta q_0, \Delta p_0, \Delta v) = -Q_2^{-2}(s_1, s_0)\int_{s_0}^{s_1}\Big[P_2(s, s_0)Q_2(s, s_0)\big(\Delta S_{11}(s) - \Delta S_{22}(s)\big)$$

$$+P_2^2(s, s_0)\Delta S_{12}(s) - Q_2^2(s, s_0)\Delta S_{21}(s)\Big]ds \tag{D.26}$$

$$= \Delta M(\Delta q_0) + \Delta M(\Delta p_0) + \Delta M(\Delta v)$$

with

$$\Delta M(\Delta q_0) = -Q_2^{-2}(s_1, s_0)\int_{s_0}^{s_1}\Big[2\frac{\partial v}{\partial q}P_2(s, s_0)\big(3Q_2(s, s_0)P_1(s, s_0) + 1\big)$$

$$-\Big(\frac{3}{v^3}\frac{\partial v}{\partial q}\frac{\partial^2 v}{\partial q^2} - \frac{1}{v^2}\frac{\partial^3 v}{\partial q^3}\Big)Q_2^2(s, s_0)Q_1(s, s_0)\Big]ds\,\Delta q_0,$$

$$\Delta M(\Delta p_0) = -Q_2^{-2}(s_1, s_0)\int_{s_0}^{s_1}\Big[6\frac{\partial v}{\partial q}P_2^2(s, s_0)Q_2(s, s_0)$$

$$-\Big(\frac{3}{v^3}\frac{\partial v}{\partial q}\frac{\partial^2 v}{\partial q^2} - \frac{1}{v^2}\frac{\partial^3 v}{\partial q^3}\Big)Q_2^3(s, s_0)\Big]ds\,\Delta p_0, \tag{D.27}$$

$$\Delta M(\Delta v) = -Q_2^{-2}(s_1, s_0)\int_{s_0}^{s_1}\Big[P_2^2(s, s_0)\Delta v(s)$$

$$-Q_2^2(s, s_0)\frac{2}{v^2}\Big(\frac{\Delta v}{v}\frac{\partial^2 v}{\partial q^2} + \frac{1}{v}\frac{\partial v}{\partial q}\frac{\partial\Delta v}{\partial q} - \frac{\Delta v}{v^2}\Big(\frac{\partial v}{\partial q}\Big)^2 - \frac{1}{2}\frac{\partial^2\Delta v}{\partial q^2}\Big)\Big]ds,$$

where, again, equations (2.30) have been used to simplify terms.

The perturbation of the ray propagator matrix due to a perturbation $\Delta s_0$ of the ray starting point along the ray is obtained from

$$\Delta\mathbf{\Pi}(\Delta s_0) = \big(\mathbf{\Pi}(s_1, s_0 + \Delta s_0) - \mathbf{\Pi}(s_1, s_0)\big) = \mathbf{\Pi}(s_1, s_0)\big(\mathbf{\Pi}^{-1}(s_0 + \Delta s_0, s_0) - \mathbf{I}_2\big). \tag{D.28}$$

The matrix $\mathbf{\Pi}(s_0 + \Delta s_0, s_0)$ is obtained from a Taylor expansion of $\mathbf{\Pi}$ around $s_0$. To first order

$$\mathbf{\Pi}(s_0 + \Delta s_0, s_0) = \mathbf{\Pi}(s_0, s_0) + \frac{d}{ds}\mathbf{\Pi}\Big|_{s_0}\Delta s_0 = \mathbf{I}_2 + \mathbf{S}\Big|_{s_0}\Delta s_0 = \begin{pmatrix} 1 & v(s_0)\Delta s_0 \\ -\frac{\Delta s_0}{v^2(s_0)}\frac{\partial^2 v}{\partial q^2} & 1 \end{pmatrix}, \tag{D.29}$$

where the fact that $\mathbf{\Pi}$ solves the paraxial ray-tracing system has been used and $\underline{\mathbf{S}}$ is defined as in equation (2.72). The inverse of matrix $\mathbf{\Pi}(s_0 + \Delta s_0, s_0)$ can be calculated using expression (2.31). Inserting the result into equation (D.28) leads to

$$\Delta\underline{\mathbf{\Pi}}(\Delta s_0) = \mathbf{\Pi}(s_1, s_0) \begin{pmatrix} 0 & -v(s_0)\Delta s_0 \\ \frac{\Delta s_0}{v^2(s_0)}\frac{\partial^2 v}{\partial q^2} & 0 \end{pmatrix} . \tag{D.30}$$

With equation (D.25) it follows that

$$\Delta M(\Delta s_0) = Q_2^{-2}(s_1, s_0)v(s_0)\Delta s_0 . \tag{D.31}$$

The results of equations (D.27) and (D.31) yield the perturbation $\Delta M$ to be used in equation (D.18). The perturbation $\Delta\alpha$ in equation (D.18) is obtained from $\Delta\alpha = v(s_1)\Delta p^{(\xi)}/\cos\alpha$ with $\Delta p^{(\xi)}$ taken from equation (D.12). Substituting for $\Delta q_0$, $\Delta p_0$ and $\Delta s_0$ from equations (D.4), (D.5), and (D.6) then yields the Fréchet derivatives of $M^{(\xi)}$:

$$\frac{\partial M_i^{(\xi)}}{\partial x_j} = -2\sin\alpha_j\cos\alpha_j\cos\theta_j\, v(s_{1j})P_1(s_{1j},s_{0j})\frac{P_2(s_{1j},s_{0j})}{Q_2(s_{1j},s_{0j})}$$
$$+\cos^2\alpha_j\Big(\frac{\partial M_j}{\partial q_{0j}}\cos\theta_j + Q_2^{-2}(s_{1j},s_{0j})v(s_{0j})\sin\theta_j\Big)\delta_{ij} ,$$

$$\frac{\partial M_i^{(\xi)}}{\partial z_j} = 2\sin\alpha_j\cos\alpha_j\sin\theta_j\, v(s_{1j})P_1(s_{1j},s_{0j})\frac{P_2(s_{1j},s_{0j})}{Q_2(s_{1j},s_{0j})}$$
$$+\cos^2\alpha_j\Big(-\frac{\partial M_j}{\partial q_{0j}}\sin\theta_j + Q_2^{-2}(s_{1j},s_{0j})v(s_{0j})\cos\theta_j\Big)\delta_{ij} , \tag{D.32}$$

$$\frac{\partial M_i^{(\xi)}}{\partial\theta_j} = -2\sin\alpha_j\cos\alpha_j\frac{v(s_{1j})}{v(s_{0j})}P_2(s_{1j},s_{0j})\frac{P_2(s_{1j},s_{0j})}{Q_2(s_{1j},s_{0j})} + \frac{1}{v(s_{0j})}\frac{\partial M_j}{\partial p_{0j}}\delta_{ij} ,$$

$$\frac{\partial M_i^{(\xi)}}{\partial v_{jk}} = -2\sin\alpha_j\cos\alpha_j\frac{P_2(s_{1j},s_{0j})}{Q_2(s_{1j},s_{0j})}\Big(P_1(s_{1j},s_{0j})\frac{\partial\tilde{q}_i}{\partial v_{jk}} + P_2(s_{1j},s_{0j})\frac{\partial\tilde{p}_i}{\partial v_{jk}}\Big) + \frac{\partial M_i}{\partial v_{jk}} ,$$

where the Kronecker symbol $\delta_{ij}$ again accounts for the fact that each data point corresponds to only one of the considered NIPs in the model and

$$\frac{\partial M_i}{\partial q_{0i}} = -Q_2^{-2}(s_{1i},s_{0i})\int_{s_{0i}}^{s_{1i}}\Big[2\frac{\partial v}{\partial q}\tilde{P}_{2i}(3\tilde{Q}_{2i}\tilde{P}_{1i}+1) - \Big(\frac{3}{v^3}\frac{\partial v}{\partial q}\frac{\partial^2 v}{\partial q^2} - \frac{1}{v^2}\frac{\partial^3 v}{\partial q^3}\Big)\tilde{Q}_{2i}^2\tilde{Q}_{1i}\Big]ds ,$$

$$\frac{\partial M_i}{\partial p_{0i}} = -Q_2^{-2}(s_{1i},s_{0i})\int_{s_{0i}}^{s_{1i}}\Big[6\frac{\partial v}{\partial q}\tilde{P}_{2i}^2\tilde{Q}_{2i} - \Big(\frac{3}{v^3}\frac{\partial v}{\partial q}\frac{\partial^2 v}{\partial q^2} - \frac{1}{v^2}\frac{\partial^3 v}{\partial q^3}\Big)\tilde{Q}_{2i}^2\Big]ds ,$$

$$\frac{\partial M_i}{\partial v_{jk}} = -Q_2^{-2}(s_{1i},s_{0i})\int_{s_{0i}}^{s_{1i}}\Big\{\Big[\tilde{P}_{2i}^2 - \tilde{Q}_{2i}^2\frac{2}{v^3}\Big(\frac{\partial^2 v}{\partial q^2} - v^{-1}\frac{\partial v}{\partial q}\frac{\partial v}{\partial q}\Big)\Big]\beta_j(x(s))\beta_k(-z(s))$$
$$+\tilde{Q}_{2i}^2\frac{1}{v^3}\Big(2\frac{\partial v}{\partial q}\frac{\partial}{\partial q}[\beta_j(x(s))\beta_k(-z(s))] - v\frac{\partial^2}{\partial q^2}[\beta_j(x(s))\beta_k(-z(s))]\Big)\Big\}ds . \tag{D.33}$$

Here, $\tilde{Q}_{1i}$, $\tilde{Q}_{2i}$, $\tilde{P}_{1i}$, and $\tilde{P}_{2i}$ are defined as $\tilde{Q}_{1i} := Q_1(s,s_{0i})$, $\tilde{Q}_{2i} := Q_2(s,s_{0i})$, $\tilde{P}_{1i} := P_1(s,s_{0i})$, and $\tilde{P}_{2i} := P_2(s,s_{0i})$.

## D.3 Fréchet derivatives of $\tau$

The traveltime along a given ray is calculated by integrating equation (2.12) along that ray:

$$\tau = \int_{s_0}^{s_1} \frac{1}{v(s)} ds . \tag{D.34}$$

From Fermat's principle, it follows that to first order, perturbations of the ray trajectory have no effect on the traveltime along the ray. This means, that for evaluating the first-order effect of perturbations of the medium velocity on the traveltime, the original ray trajectory can be used. Perturbations of the ray starting and end points perpendicular to the original ray as well as perturbations of the initial ray direction for fixed ray starting and end points have no first order effect on traveltime, see equation (2.67). Consequently, traveltime perturbations due to perturbations of the velocity model and the NIP model parameters may be written as

$$
\begin{aligned}
\Delta\tau &= \int_{s_0}^{s_1} \Delta\left(\frac{1}{v(s)}\right) ds - \frac{\Delta s_0}{v(s_0)} \\
&= -\int_{s_0}^{s_1} \frac{\Delta v(s)}{v^2(s)} ds - \frac{\sin\theta}{v(s_0)}\Delta x - \frac{\cos\theta}{v(s_0)}\Delta z ,
\end{aligned}
\tag{D.35}
$$

where the integration is taken along the original ray. Again making use of the B-spline description of the velocity model and of the fact that the integral in equation (D.35) is a linear functional in $\Delta v$, the Fréchet derivatives of the traveltime $\tau$ can be written as follows:

$$
\begin{aligned}
\frac{\partial\tau_i}{\partial x_j} &= -\frac{\sin\theta_j}{v(s_{0j})}\delta_{ij} , \\
\frac{\partial\tau_i}{\partial z_j} &= -\frac{\cos\theta_j}{v(s_{0j})}\delta_{ij} , \\
\frac{\partial\tau_i}{\partial\theta_j} &= 0 , \\
\frac{\partial\tau_i}{\partial v_{jk}} &= -\int_{s_{0i}}^{s_{1i}} \frac{\beta_j(x(s))\beta_k(-z(s))}{v^2(s)} ds .
\end{aligned}
\tag{D.36}
$$

All Fréchet derivatives derived in this appendix can be computed during ray tracing along each of the considered normal rays. For that purpose, the integrals in equations (D.15), (D.16), (D.33), and (D.36) are evaluated numerically along each ray. Wherever perturbations of the velocity are involved, the integration needs to be performed separately for each of the $n_x n_z$ velocity model parameters $v_{jk}$.

The drawback of performing ray tracing and the calculation of Fréchet derivatives in ray-centered coordinates lies in the fact that evaluations of spatial velocity derivatives in arbitrary directions (defined by the local ray normal direction) are required. While velocity derivatives in the $x$ and $z$ directions can be calculated using equation (B.8), spatial derivatives in any other spatial direction require additional B-spline evaluations, depending on the order of the derivative. As the integrals in equations (D.15), (D.16), and (D.33) contain velocity derivatives up to third order, and derivatives of velocity perturbations need to be evaluated separately for each velocity model parameter $v_{jk}$, this results in significantly increased computation times.

## D.4  Fréchet derivatives for 1D tomographic inversion

In the 1D case, velocities do not vary in the horizontal direction and all normal rays are vertical. The data components involved in 1D tomographic inversion will here be denoted by $(\tau, M)$, while each NIP in the subsurface is simply characterized by its depth, denoted here by $z$. Note that $z < 0$, as all NIPs are located below the measurement surface. The velocity model parameters are the B-spline coefficients in equation (5.3). The 1D Fréchet derivatives can be obtained as special cases of the 2D expressions given above. As in the 1D case there is no horizontal velocity variation, the paraxial ray-tracing system simplifies considerably. The velocity derivative in matrix $\underline{\mathbf{S}}$, equation (2.72), is zero, resulting in closed-form expressions for the elements of the ray propagator matrix:

$$\underline{\mathbf{\Pi}}(0,z) = \begin{pmatrix} 1 & \int_z^0 v(z')\,dz' \\ 0 & 1 \end{pmatrix} . \tag{D.37}$$

Consequently, the Fréchet derivatives of $M$ are

$$\frac{\partial M_i}{\partial z_j} = v(z_j) \left[ \int_{z_j}^0 v(z')\,dz' \right]^{-2} \delta_{ij} ,$$

$$\frac{\partial M_i}{\partial v_k} = - \int_{z_i}^0 \beta_k(-z')\,dz' \left[ \int_{z_i}^0 v(z')\,dz' \right]^{-2} , \tag{D.38}$$

while the Fréchet derivatives of $\tau$ follow from equations (D.36):

$$\frac{\partial \tau_i}{\partial z_j} = -\frac{1}{v(z_j)} \delta_{ij} ,$$

$$\frac{\partial \tau_i}{\partial v_k} = - \int_{z_i}^0 \frac{\beta_k(-z')}{v^2(z')}\,dz' . \tag{D.39}$$

# Appendix E

# Fréchet derivatives for 3D tomographic inversion

For the 3D tomographic inversion described in Section 5.3, the Fréchet derivatives of the quantities $(\tau_0, M_\phi^{(\xi)}, p_x^{(\xi)}, p_y^{(\xi)}, \xi_x, \xi_y)_i$ with respect to the NIP model parameters $(x, y, z, e_x, e_y)_i^{(NIP)}$ and the velocity model parameters $v_{jkl}$ are required. Compared to the 2D Fréchet derivatives (Appendix D), the expressions for the Fréchet derivatives in 3D are significantly more complicated. This is due not only to the additional dimension, but also to the use of a different coordinate system: while in the 2D case many of the involved calculations simplify because of the formulation in ray-centered coordinates, this is not the case in 3D, where global Cartesian coordinates with the $z$-coordinate as the independent parameter along the ray (Section 2.6) are used. Therefore, only a brief outline of the derivation of the 3D Fréchet derivatives is given in this appendix. The resulting expressions will not be written out explicitly.

As in the 2D case, a simplified notation for the model and data components is used. Spatial coordinates will be denoted by $x$, $y$, and $z$, while the corresponding slowness components are written as $p_x$, $p_y$, and $p_z$. For quantities at the ray starting point (the NIP), a subscript 0 is used. A NIP is, thus, characterized by $(x_0, y_0, z_0, e_{x0}, e_{y0})$. For quantities at the ray endpoint (at the measurement surface) a subscript 1 is used. The slowness components and coordinates $p_x^{(\xi)}, p_y^{(\xi)}, \xi_x$, and $\xi_y$ used as data components in the inversion are therefore identical to $p_{x1}, p_{y1}, x_1$, and $y_1$.

## E.1 Fréchet derivatives of $\xi_x$, $\xi_y$, $p_x^{(\xi)}$, and $p_y^{(\xi)}$

According to Section 2.4, perturbations $\Delta\boldsymbol{\eta}_1 = (\Delta x_1, \Delta y_1, \Delta p_{x1}, \Delta p_{y1})$ of the coordinates and slowness components at the ray endpoint can be related to perturbations $\Delta\boldsymbol{\eta}_0$ of the corresponding quantities at the ray starting point and perturbations $\Delta v$ of the velocity along the ray by

$$\Delta\boldsymbol{\eta}_1 = \underline{\underline{\Pi}}^{(x)}(z_1, z_0)\,\Delta\boldsymbol{\eta}_0 + \int_{z_0}^{z_1} \underline{\underline{\Pi}}^{(x)-1}(z, z_0)\,\Delta\mathbf{w}(z, \Delta v(z))\,dz \,, \qquad (E.1)$$

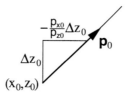

Figure E.1: Geometrical relation between perturbations of the vertical and the horizontal coordinates of the ray starting point.

where $\underline{\mathbf{\Pi}}^{(x)}$ is the ray propagator matrix associated with the reduced paraxial ray-tracing system in global Cartesian coordinates introduced in Section 2.6. The elements of $\underline{\mathbf{\Pi}}^{(x)-1}(z, z_0)$ can be obtained from the elements of $\underline{\mathbf{\Pi}}^{(x)}(z, z_0)$ via equation (2.31). The vector $\Delta\mathbf{w}$ is defined in equation (2.44). With the definition of $H$ given in equation (2.78) and with

$$\Delta H = \frac{\partial H}{\partial v} \Delta v = \frac{1}{v^3 p_z} \Delta v = \frac{1}{v^3 \sqrt{v^{-2} - p_x^2 - p_y^2}} \Delta v \ , \tag{E.2}$$

it may be written as

$$
\begin{aligned}
\Delta\mathbf{w} &= \left( \frac{\partial}{\partial p_x} \Delta H, \frac{\partial}{\partial p_y} \Delta H, -\frac{\partial}{\partial x} \Delta H, -\frac{\partial}{\partial y} \Delta H \right)^T \\
&= \frac{1}{v^3 p_z^3} \begin{pmatrix} p_x \, \Delta v \\ p_y \, \Delta v \\ -\left(\frac{1}{v^3} - \frac{3 p_z^2}{v}\right) \frac{\partial v}{\partial x} \Delta v - p_z^2 \frac{\partial \Delta v}{\partial x} \\ -\left(\frac{1}{v^3} - \frac{3 p_z^2}{v}\right) \frac{\partial v}{\partial y} \Delta v - p_z^2 \frac{\partial \Delta v}{\partial y} \end{pmatrix} \ .
\end{aligned}
\tag{E.3}
$$

Making use of the generalization of equation (D.13) to the 3D case and taking into account that $e_{x0} = p_{x0} v(x_0, y_0, z_0)$ and $e_{y0} = p_{y0} v(x_0, y_0, z_0)$, equation (E.1) yields the Fréchet derivatives of the data components $(\xi_x, \xi_y, p_x^{(\xi)}, p_y^{(\xi)}) \equiv (x_1, y_1, p_{x1}, p_{y1})$ with respect to the NIP model parameters $(x_0, y_0, e_{x0}, e_{y0})$ of the same ray and with respect to the velocity model parameters $v_{jkl}$.

The Fréchet derivatives of $\xi_x$, $\xi_y$, $p_x^{(\xi)}$, and $p_y^{(\xi)}$ with respect to perturbations $\Delta z_0$ of the $z$-coordinate of the ray starting point can be obtained from

$$
\begin{aligned}
\Delta\boldsymbol{\eta}_1(\Delta z_0) &= \underline{\mathbf{\Pi}}^{(x)}(z_1, z_0) \, \Delta\boldsymbol{\eta}_0(\Delta z_0) \\
&= \underline{\mathbf{\Pi}}^{(x)}(z_1, z_0) \begin{pmatrix} -\frac{p_{x0}}{p_{z0}} \Delta z_0 \\ -\frac{p_{y0}}{p_{z0}} \Delta z_0 \\ 0 \\ 0 \end{pmatrix} \ ,
\end{aligned}
\tag{E.4}
$$

where the elements of $\Delta\boldsymbol{\eta}_0(\Delta z_0)$ are obtained from geometrical considerations (see Figure E.1).

## E.2 Fréchet derivatives of $M_\phi^{(\xi)}$

The matrix $\underline{\mathbf{M}}_{\mathrm{NIP}}^{(\xi)}$ can be calculated from the elements of the ray propagator matrix $\underline{\boldsymbol{\Pi}}^{(x)}$, equation (2.82), via equation (5.42). Perturbations of $\underline{\mathbf{M}}_{\mathrm{NIP}}^{(\xi)}$ can be written in terms of perturbations of the elements of $\underline{\boldsymbol{\Pi}}^{(x)}$ as

$$\Delta\underline{\mathbf{M}}_{\mathrm{NIP}}^{(\xi)} = \left(\Delta\underline{\mathbf{P}}_2^{(x)} - \underline{\mathbf{P}}_2^{(x)}\underline{\mathbf{Q}}_2^{(x)-1}\Delta\underline{\mathbf{Q}}_2^{(x)}\right)\underline{\mathbf{Q}}_2^{(x)-1} . \tag{E.5}$$

Perturbations of $M_\phi^{(\xi)}$ defined in equation (5.43) for a given value of $\phi$ then read

$$\Delta M_\phi^{(\xi)} = \cos^2\phi\,\left(\Delta M_{\mathrm{NIP}}^{(\xi)}\right)_{11} + 2\sin\phi\cos\phi\,\left(\Delta M_{\mathrm{NIP}}^{(\xi)}\right)_{12} + \sin^2\phi\,\left(\Delta M_{\mathrm{NIP}}^{(\xi)}\right)_{22} , \tag{E.6}$$

where $\left(\Delta M_{\mathrm{NIP}}^{(\xi)}\right)_{11}$, $\left(\Delta M_{\mathrm{NIP}}^{(\xi)}\right)_{12}$, and $\left(\Delta M_{\mathrm{NIP}}^{(\xi)}\right)_{22}$ are the elements of $\Delta\underline{\mathbf{M}}^{(\xi)}$. Perturbations of the ray propagator matrix $\underline{\boldsymbol{\Pi}}^{(x)}$ itself can, in turn, be related to perturbations of the horizontal coordinates and associated slowness components of the ray starting point $(x_0, y_0, p_{x0}, p_{y0})$ and perturbations $\Delta v$ of the velocity along the ray with the help of equation (2.54):

$$\Delta\underline{\boldsymbol{\Pi}}^{(x)} = \underline{\boldsymbol{\Pi}}^{(x)}(z_1, z_0)\int_{z_0}^{z_1}\underline{\boldsymbol{\Pi}}^{(x)-1}(z, z_0)\Delta\underline{\mathbf{S}}(z)\underline{\boldsymbol{\Pi}}^{(x)}(z, z_0)\,dz , \tag{E.7}$$

where

$$\Delta\underline{\mathbf{S}}(z) = \Delta\underline{\mathbf{S}}_1\left(\Delta v(z)\right) + \Delta\underline{\mathbf{S}}_2\left(\Delta x_0, \Delta y_0, \Delta p_{x0}, \Delta p_{y0}\right) . \tag{E.8}$$

The matrices $\Delta\underline{\mathbf{S}}_1$ and $\Delta\underline{\mathbf{S}}_2$ are defined in equations (2.49) and (2.50). Their elements can be calculated from the Hamiltonian $H$, equation (2.78), and the perturbed Hamiltonian given in equation (E.2). The resulting expressions are lengthy and will not be written out explicitly in this appendix. The quantities $\Delta x(z)$, $\Delta y(z)$, $\Delta p_x(z)$, and $\Delta p_y(z)$ required for the calculation of $\Delta\underline{\mathbf{S}}_2$ along the ray can be obtained from the corresponding quantities at the ray starting point with equation (E.1). For practical purposes, the dependence of $\Delta\underline{\mathbf{S}}_2$ on perturbations of the velocity along the ray via the second term in equation (E.1) will, however, be neglected.

Analogously to equations (D.24) and (D.25) in the 2D case, it can be shown that if perturbations of the ray propagator matrix $\underline{\boldsymbol{\Pi}}^{(x)}$ can be written in the form

$$\Delta\underline{\boldsymbol{\Pi}}^{(x)} = \underline{\boldsymbol{\Pi}}^{(x)}\mathbf{A} = \begin{pmatrix}\underline{\mathbf{Q}}_1^{(x)} & \underline{\mathbf{Q}}_2^{(x)} \\ \underline{\mathbf{P}}_1^{(x)} & \underline{\mathbf{P}}_2^{(x)}\end{pmatrix}\begin{pmatrix}\underline{\mathbf{A}}^{(1)} & \underline{\mathbf{A}}^{(2)} \\ \underline{\mathbf{A}}^{(3)} & \underline{\mathbf{A}}^{(4)}\end{pmatrix} , \tag{E.9}$$

where $\underline{\mathbf{A}}^{(1)}$, $\underline{\mathbf{A}}^{(2)}$, $\underline{\mathbf{A}}^{(3)}$, and $\underline{\mathbf{A}}^{(4)}$ are $2 \times 2$ submatrices of $\underline{\mathbf{A}}$, the perturbation $\Delta\underline{\mathbf{M}}_{\mathrm{NIP}}^{(\xi)}$, equation (E.5), reduces to

$$\Delta\underline{\mathbf{M}}_{\mathrm{NIP}}^{(\xi)} = -\left(\underline{\mathbf{Q}}_2^{(x)-1}\right)^T\underline{\mathbf{A}}^{(2)}\underline{\mathbf{Q}}_2^{(x)-1} . \tag{E.10}$$

Consequently, only the upper right $2 \times 2$ submatrix of the matrix under the integral in equation (E.7) is required. Making use of equation (2.31) to express $\underline{\boldsymbol{\Pi}}^{(x)-1}(z, z_0)$ in terms of the elements of $\underline{\boldsymbol{\Pi}}^{(x)}(z, z_0)$ and using the notation

$$\Delta\underline{\mathbf{S}} = \Delta\underline{\mathbf{S}}_1 + \Delta\underline{\mathbf{S}}_2 = \begin{pmatrix}\Delta\underline{\mathbf{S}}^{(1)} & \Delta\underline{\mathbf{S}}^{(2)} \\ \Delta\underline{\mathbf{S}}^{(3)} & \Delta\underline{\mathbf{S}}^{(4)}\end{pmatrix} , \tag{E.11}$$

where $\Delta\underline{S}^{(1)}$, $\Delta\underline{S}^{(2)}$, $\Delta\underline{S}^{(3)}$, and $\Delta\underline{S}^{(4)}$ are $2 \times 2$ submatrices of $\Delta\underline{S}$, the perturbation $\Delta\underline{M}^{(\xi)}_{\mathrm{NIP}}$ due to perturbations of the quantities $x_0$, $y_0$, $p_{x0}$, and $p_{y0}$ and perturbations of the velocity along the ray can be expressed as

$$\Delta\underline{M}^{(\xi)}_{\mathrm{NIP}} = -\left(\underline{Q}^{(x)-1}_2\right)^T \int_{z_0}^{z_1} \left\{ \underline{P}^{(x)T}_2(z,z_0)\left[\Delta\underline{S}^{(1)}\underline{Q}^{(x)}_2(z,z_0) + \Delta\underline{S}^{(2)}\underline{P}^{(x)}_2(z,z_0)\right] \right.$$
$$\left. - \underline{Q}^{(x)T}_2(z,z_0)\left[\Delta\underline{S}^{(3)}\underline{Q}^{(x)}_2(z,z_0) + \Delta\underline{S}^{(4)}\underline{P}^{(x)}_2(z,z_0)\right] \right\} dz\,\underline{Q}^{(x)-1}_2 \ . \tag{E.12}$$

The Fréchet derivatives of $M^{(\xi)}_\phi$ with respect to the NIP model parameters $(x_0, y_0, e_{x0}, e_{y0})$ and with respect to the velocity model parameters $v_{jkl}$ can be obtained by substituting the correspondingg components of $\Delta\underline{S}$, equation (E.8), into equation (E.12) and using the resulting components of $\Delta\underline{M}^{(\xi)}_{\mathrm{NIP}}$ in equation (E.6). Again, equation (D.13) (generalized to the 3D case) needs to be applied to obtain the derivatives with respect to the velocity parameters $v_{jkl}$.

Perturbations of the ray propagator matrix $\underline{\Pi}^{(x)}(z_1, z_0)$ due to a perturbation $\Delta z_0$ of the ray starting point in the vertical direction can be obtained from

$$\Delta\underline{\Pi}^{(x)}(\Delta z_0) = \left(\underline{\Pi}^{(x)}(z_1, z_0 + \Delta z_0) - \underline{\Pi}^{(x)}(z_1, z_0)\right)$$
$$= \underline{\Pi}^{(x)}(z_1, z_0)\left(\underline{\Pi}^{(x)-1}(z_0 + \Delta z_0, z_0) - \underline{I}_4\right) \ . \tag{E.13}$$

A Taylor expansion of $\underline{\Pi}^{(x)}$ around $z_0$ to first order yields

$$\underline{\Pi}^{(x)}(z_0 + \Delta z_0, z_0) \approx \underline{\Pi}^{(x)}(z_0, z_0) + \frac{d}{dz}\underline{\Pi}^{(x)}\Big|_{z_0}\Delta z_0 = \underline{I}_4 + \underline{S}\Big|_{z_0}\Delta z_0 \ , \tag{E.14}$$

where the matrix $\underline{S}$ has been defined in Section 2.6. Expressing the inverse of the propagator matrix $\underline{\Pi}^{(x)}(z_0 + \Delta z_0, z_0)$ in terms of its $2 \times 2$ submatrices and substituting it into equation (E.13) yields

$$\Delta\underline{\Pi}^{(x)}(\Delta z_0) = -\underline{\Pi}^{(x)}(z_1, z_0)\underline{S}\Big|_{z_0}\Delta z_0 \ . \tag{E.15}$$

According to equation (E.10), the perturbation $\Delta\underline{M}^{(\xi)}_{\mathrm{NIP}}$ due to a perturbation $\Delta z_0$ can, thus, be expressed as

$$\Delta\underline{M}^{(\xi)}_{\mathrm{NIP}} = \left(\underline{Q}^{(x)-1}_2\right)^T \underline{S}^{(2)}\Big|_{z_0}\underline{Q}^{(x)-1}_2\Delta z_0 \ , \tag{E.16}$$

where $\underline{S}^{(2)}$ is the upper right $2 \times 2$ submatrix of $\underline{S}$, given by

$$\underline{S}^{(2)} = \begin{pmatrix} \frac{1}{p_z} + \frac{p_x^2}{p_z^3} & \frac{p_x p_y}{p_z^3} \\ \frac{p_x p_y}{p_z^3} & \frac{1}{p_z} + \frac{p_y^2}{p_z^3} \end{pmatrix} \ . \tag{E.17}$$

Again, the components of the resulting matrix $\Delta\underline{M}^{(\xi)}_{\mathrm{NIP}}(\Delta z_0)$ need to be used in equation (E.6) to obtain the Fréchet derivatives of $M^{(\xi)}_\phi$ with respect to the coordinate $z_0$ of the considered ray starting point.

# E.3 Fréchet derivatives of $\tau_0$

As in the 2D case, for the computation of the Fréchet derivatives of the traveltime $\tau_0$ Fermat's principle can be applied. It says that perturbations of the ray trajectory have no first-order effect on the traveltime along the ray. It follows that for evaluating the effect of perturbations of the velocity along the ray on the traveltime, the original ray trajectory can be used. The effects of velocity perturbations on the ray trajectory can be neglected. Thus,

$$\Delta\tau_0(\Delta v) = \int_{s_0}^{s_1} \Delta\left(\frac{1}{v}\right) ds = -\int_{s_0}^{s_1} \frac{\Delta v}{v^2} ds = -\int_{z_0}^{z_1} \frac{\Delta v}{v^3 p_z} dz , \qquad \text{(E.18)}$$

where $s$ is the arclength along the ray and

$$\begin{aligned}
ds &= \sqrt{dx^2 + dy^2 + dz^2} \\
&= \sqrt{\left[\left(\frac{dx}{dz}\right)^2 + \left(\frac{dy}{dz}\right)^2 + 1\right]} \, dz \\
&= \sqrt{p_x^2 + p_y^2 + p_z^2} \, \frac{dz}{p_z} \\
&= \frac{1}{v p_z} \, dz
\end{aligned} \qquad \text{(E.19)}$$

has been used. In order to obtain the derivatives with respect to the velocity model parameters $v_{jkl}$, the generalization of equation (D.13) to the 3D case again needs to be applied.

Also due to Fermat's principle, perturbations of the initial slowness components of the ray have no first-order effect on the traveltime. The corresponding Fréchet derivatives are zero. The same is true for perturbations of the ray starting and end points in directions perpendicular to the ray. Only perturbations of the ray starting and end points along the ray direction have an effect on the traveltime. Thus,

$$\Delta\tau_0(\Delta x_0, \Delta y_0, \Delta z_0, \Delta p_{x0}, \Delta p_{y0}) = -p_{x0}\Delta x_0 - p_{y0}\Delta y_0 - p_{z0}\Delta z_0 , \qquad \text{(E.20)}$$

which immediately yields the required Fréchet derivatives of $\tau_0$ with respect to the NIP model parameters $x_0$, $y_0$, $z_0$, $e_{x0}$, and $e_{y0}$.

# Appendix F

# Fréchet derivatives for additional model constraints

The additional model constraints discussed in Section 4.5 are in the context of the tomographic inversion treated as extra data. In order to minimize the misfit between these data, given in the 2D case by equation (5.31) and in the 3D case by equation (5.46), and the corresponding modeled quantities during the inversion, the associated Fréchet derivatives with respect to the model parameters are required. In this appendix, these derivatives, to be used in the tomographic matrix, equation (5.31) or (5.46), respectively, are derived.

## F.1  A priori velocity information

A priori information on seismic velocities is in the 2D case given in the form of velocity values specified at $n_{\text{vdata}}$ spatial locations in the model. The corresponding modeled quantities in a given velocity model are

$$v(x_i, z_i) = \sum_{j=1}^{n_x} \sum_{k=1}^{n_z} v_{jk} \beta_j(x_i) \beta_k(-z_i) \qquad i = 1, \ldots, n_{\text{vdata}} . \tag{F.1}$$

This expression obviously depends only on the velocity model parameters $v_{jk}$ and not on any of the NIP model parameters. The Fréchet derivatives with respect to the parameters $v_{jk}$ follow directly from equation (F.1):

$$\frac{\partial v(x_i, z_i)}{\partial v_{jk}} = \beta_j(x_i) \beta_k(-z_i) . \tag{F.2}$$

These quantities are the elements of the submatrix $\left[\frac{\partial v^{(\text{constr})}}{\partial v}\right]$ of $\hat{\mathbf{F}}$, equation (5.35).

Analogously, in the 3D case the modeled quantities corresponding to the a priori velocity information specified at $n_{\text{vdata}}$ spatial locations in the model are

$$v(x_i, y_i, z_i) = \sum_{j=1}^{n_x} \sum_{k=1}^{n_y} \sum_{l=1}^{n_z} v_{jkl} \beta_j(x_i) \beta_k(y_i) \beta_l(-z_i) \qquad i = 1, \ldots, n_{\text{vdata}} . \tag{F.3}$$

The corresponding Fréchet derivatives with respect to the velocity model parameters are simply

$$\frac{\partial v(x_i, y_i, z_i)}{\partial v_{jkl}} = \beta_j(x_i)\beta_k(y_i)\beta_l(-z_i) \, . \tag{F.4}$$

These are used as the elements of the submatrix $\left[\frac{\partial v^{(\text{constr})}}{\partial v}\right]$ of $\hat{\mathbf{F}}$, equation (5.50).

## F.2   Minimum local velocity variation along reflectors

As described in Section 4.5, the constraint that the velocity structure should locally follow the reflector structure is imposed by locally minimizing the velocity gradient along the reflector at each considered NIP. This local velocity gradient obviously depends on the velocity distribution, but also on the location at which it is evaluated and on the local reflector orientation. It is, thus, dependent on the NIP model parameters. However, as briefly discussed in Section 5.2, the NIP locations and associated local reflector orientations usually do not vary drastically from iteration to iteration. Therefore, the derivatives of the local velocity gradient along the reflector at each NIP location with respect to the NIP model parameters are neglected, that is, assumed to be zero. This significantly simplifies the implementation of the constraint and makes the inversion more efficient.

In the 2D case, the velocity gradient along the reflector at a given NIP location simply consists of the local spatial derivative of velocity in the direction perpendicular to the normal ray at the NIP. In accordance with the notation of Section 2.5, the spatial coordinate associated with that direction is denoted by $q$. The velocity derivative can be written as

$$
\begin{aligned}
\left(\nabla_q v\right)\Big|_{(x,z)_i^{(\text{NIP})}} &= \frac{\partial v}{\partial q}\Big|_{(x,z)_i^{(\text{NIP})}} \\
&= \cos\theta_i^{(\text{NIP})} \frac{\partial v}{\partial x}\Big|_{(x,z)_i^{(\text{NIP})}} - \sin\theta_i^{(\text{NIP})} \frac{\partial v}{\partial z}\Big|_{(x,z)_i^{(\text{NIP})}} \\
&= \cos\theta_i^{(\text{NIP})} \sum_{j=1}^{n_x}\sum_{k=1}^{n_z} v_{jk}\left(\frac{\partial\beta_j(x)}{\partial x}\beta_k(-z)\right)\Big|_{(x,z)_i^{(\text{NIP})}} \\
&\quad + \sin\theta_i^{(\text{NIP})} \sum_{j=1}^{n_x}\sum_{k=1}^{n_z} v_{jk}\left(\beta_j(x)\frac{\partial\beta_k(-z)}{\partial z}\right)\Big|_{(x,z)_i^{(\text{NIP})}} \, .
\end{aligned} \tag{F.5}
$$

Accordingly, the Fréchet derivatives with respect to the velocity model parameters $v_{jk}$ are

$$
\begin{aligned}
\frac{\partial}{\partial v_{jk}}\left[\left(\nabla_q v\right)\Big|_{(x,z)_i^{(\text{NIP})}}\right] &= \cos\theta_i^{(\text{NIP})}\left(\frac{\partial\beta_j(x)}{\partial x}\beta_k(-z)\right)\Big|_{(x,z)_i^{(\text{NIP})}} \\
&\quad + \sin\theta_i^{(\text{NIP})}\left(\beta_j(x)\frac{\partial\beta_k(-z)}{\partial z}\right)\Big|_{(x,z)_i^{(\text{NIP})}} \, .
\end{aligned} \tag{F.6}
$$

If, instead of $\nabla_q v$, its absolute value $|\nabla_q v|$ is considered, the Fréchet derivatives become

$$\frac{\partial}{\partial v_{jk}}\left[\left(|\nabla_q v|\right)\Big|_{(x,z)_i^{(\text{NIP})}}\right] = \text{sgn}\left(\left(\nabla_q v\right)\Big|_{(x,z)_i^{(\text{NIP})}}\right)\frac{\partial}{\partial v_{jk}}\left[\left(\nabla_q v\right)\Big|_{(x,z)_i^{(\text{NIP})}}\right] \, , \tag{F.7}$$

where the sgn-function takes the values $+1$, $-1$, or $0$, depending on the sign of its argument. The Fréchet derivatives in equation (F.7) are the elements of the submatrix $\left[\frac{\partial(\nabla_{\mathbf{q}}v)}{\partial v}\right]$ of $\hat{\mathbf{F}}$, equation (5.35).

In the 3D case, the two unit vectors $\hat{\mathbf{e}}_1$ and $\hat{\mathbf{e}}_2$ defined in Section 4.5 span the plane that is locally tangent to the reflector at a given NIP location. The local velocity gradient in the tangent plane to the reflector at the considered NIP is then defined by

$$
(\nabla_{\mathbf{q}}v)\Big|_{(x,y,z)_i^{(\mathrm{NIP})}} = (\hat{\mathbf{e}}_1 \cdot \nabla v, \hat{\mathbf{e}}_2 \cdot \nabla v)^T \Big|_{(x,y,z)_i^{(\mathrm{NIP})}} = \left(\frac{\partial v}{\partial q_1}, \frac{\partial v}{\partial q_2}\right)^T \Big|_{(x,y,z)_i^{(\mathrm{NIP})}} \tag{F.8}
$$

where the vector $\mathbf{q}$ contains the ray-centered coordinates $q_1$ and $q_2$ defined in Section 5.2. The magnitude of this velocity gradient is

$$
|\nabla_{\mathbf{q}}v|\Big|_{(x,y,z)_i^{(\mathrm{NIP})}} = \left[\left(\frac{\partial v}{\partial q_1}\right)^2 + \left(\frac{\partial v}{\partial q_2}\right)^2\right]^{1/2} \Big|_{(x,y,z)_i^{(\mathrm{NIP})}} \tag{F.9}
$$

with

$$
\begin{aligned}
\frac{\partial v}{\partial q_1} = \hat{\mathbf{e}}_1 \cdot \nabla v = \sum_{j=1}^{n_x}\sum_{k=1}^{n_y}\sum_{l=1}^{n_z} v_{jkl}\left(e_{1x}\frac{\partial \beta_j(x)}{\partial x}\beta_k(y)\beta_l(-z)\right. \\
\left. + e_{1y}\beta_j(x)\frac{\partial \beta_k(y)}{\partial y}\beta_l(-z) - e_{1z}\beta_j(x)\beta_k(y)\frac{\partial \beta_l(-z)}{\partial z}\right)
\end{aligned} \tag{F.10}
$$

and an analogous expression for $\frac{\partial v}{\partial q_2} = \hat{\mathbf{e}}_2 \cdot \nabla v$. The Fréchet derivatives of expression (F.9) with respect to the velocity model parameters $v_{jkl}$ read

$$
\frac{\partial}{\partial v_{jkl}}\left[|\nabla_{\mathbf{q}}v|\Big|_{(x,y,z)_i^{(\mathrm{NIP})}}\right] = \left\{\frac{1}{|\nabla_{\mathbf{q}}v|}\left[\frac{\partial v}{\partial q_1}\frac{\partial}{\partial v_{jkl}}\left(\frac{\partial v}{\partial q_1}\right) + \frac{\partial v}{\partial q_2}\frac{\partial}{\partial v_{jkl}}\left(\frac{\partial v}{\partial q_2}\right)\right]\right\}\Big|_{(x,y,z)_i^{(\mathrm{NIP})}}, \tag{F.11}
$$

where $\frac{\partial v}{\partial q_1}$ is given by equation (F.10) and

$$
\frac{\partial}{\partial v_{jkl}}\left(\frac{\partial v}{\partial q_1}\right) = \left(e_{1x}\frac{\partial \beta_j(x)}{\partial x}\beta_k(y)\beta_l(-z) + e_{1y}\beta_j(x)\frac{\partial \beta_k(y)}{\partial y}\beta_l(-z) - e_{1z}\beta_j(x)\beta_k(y)\frac{\partial \beta_l(-z)}{\partial z}\right). \tag{F.12}
$$

The expression for $\frac{\partial}{\partial v_{jkl}}\left(\frac{\partial v}{\partial q_2}\right)$ has the same form as equation (F.12). The Fréchet derivatives (F.11) are the elements of the submatrix $\left[\frac{\partial(\nabla_{\mathbf{q}}v)}{\partial v}\right]$ of $\hat{\mathbf{F}}$, equation (5.50).

# Appendix G

# Smoothing of kinematic wavefield attributes

During the CRS stack procedure, a set of kinematic wavefield attributes is determined separately for each zero-offset sample. Wavefield attributes determined in this way may fluctuate from sample to sample on a reflection event due to noise in the seismic data. Also, a search strategy as described by Mann et al. (1999), in which the attributes are determined with one-parameter searches in subsets of the data, may fail to detect the global coherence maximum for the entire CRS operator. This can lead to outliers in the determined attributes.

Fluctuations in the kinematic wavefield attributes, as well as the presence of outliers may have adverse effects on the stack result, but will also degrade the performance of the attribute-based tomographic inversion and other attribute-based processes. In this appendix, a simple, but effective smoothing algorithm is presented, which removes such fluctuations and outliers in an event-consistent manner. For simplicity, only the 2D case will be treated here. An extension to 3D is straightforward.

If paraxial ray theory is valid in the vicinity of each zero-offset ray associated with a reflection event, as is assumed during the CRS stack, the kinematic wavefield attributes $p^{(\xi)}$, $M_{\text{NIP}}^{(\xi)}$, and $M_{\text{N}}^{(\xi)}$ should theoretically vary smoothly along reflection events. Also—because they represent spatial traveltime derivatives—these quantities should remain constant along the reflection signal in the temporal direction, as the time-domain signal length of a reflection event is spatially invariant in unmigrated data. From equation (3.8) it follows that if these statements are true for $p^{(\xi)}$, $M_{\text{NIP}}^{(\xi)}$, and $M_{\text{N}}^{(\xi)}$, they also hold for $\alpha$, $K_{\text{NIP}}$, and $K_{\text{N}}$, and, in fact, also for $R_{\text{NIP}}$. All deviations from this behavior in the attributes determined with the CRS stack can therefore be assumed to be due to the above-mentioned effects of noise in the seismic data and shortcomings of the search strategy.

This justifies the application of a local smoothing filter to the kinematic wavefield attributes along reflection events to remove outliers and fluctuations. An appropriate smoothing algorithm should yield physically meaningful attribute values without destroying any relevant information. It should work in an event-consistent way, taking into account the local reflector dip. Attributes from different reflection events, characterized by different local reflector dips, should not be mixed. The

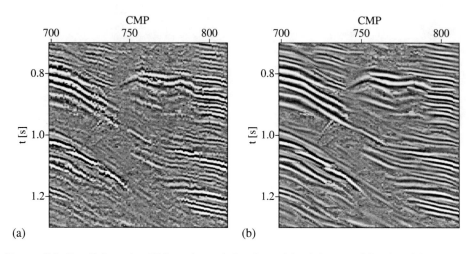

Figure G.1: Detail from the CRS stack result for the real land dataset of Section 6.3. (a) Stack with the original attributes obtained from one-parameter searches. (b) Stack result with smoothed $\alpha$ and $R_{\mathrm{NIP}}$ sections.

relevant dip information is available for each zero-offset sample from

$$\frac{\partial t_0}{\partial \xi} = 2\,p^{(\xi)} = \frac{2\sin\alpha}{v_0} \ . \tag{G.1}$$

It can be determined from the seismic data in a stable way and usually exhibits little fluctuation on an event. It can therefore be used to reliably differentiate between zero-offset samples from different reflection events on grounds of their differing dip values. Further, only attributes associated with sufficiently high coherence values should be used. A low coherence indicates that the kinematic wavefield attributes are unreliable or that the considered zero-offset sample is not located on a reflection event.

In order to handle fluctuations, as well as outliers in the kinematic wavefield attributes, the smoothing algorithm described here makes use of a combination of averaging and a median filter. It involves the following steps:

- Define a smoothing window with a temporal extension of $n_t$ samples and a spatial extension of $n_\xi$ traces, centered around the considered zero-offset sample.

- Orient the window along the reflection event using the dip information, given by equation (G.1), associated with the central zero-offset sample.

- Reject all attributes in the window with coherence values below a specified threshold.

- Reject all attributes in the window for which the angle $\alpha$ deviates from the angle at the central zero-offset sample by more than a specified value $\Delta\alpha$.

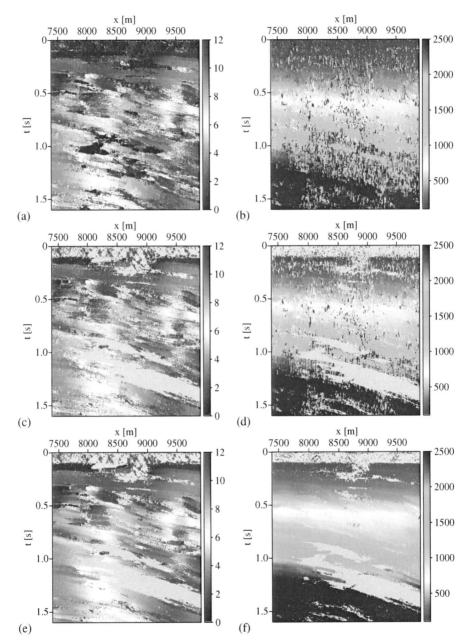

Figure G.2: Smoothing of CRS attribute sections, real data example with low S/N ratio. (a) Original $\alpha$ section. (b) Original $R_{\mathrm{NIP}}$ section. (c) Original $\alpha$ section with samples corresponding to low coherence values masked. (d) Original $R_{\mathrm{NIP}}$ section, masked. (e) Smoothed $\alpha$ section, masked. (f) Smoothed $R_{\mathrm{NIP}}$ section, masked.

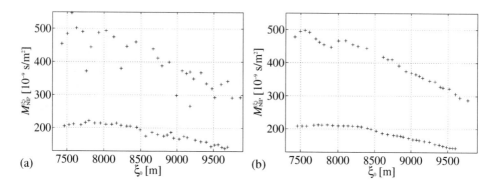

Figure G.3: (a) $M_{\mathrm{NIP}}^{(\xi)}$ values extracted from the unsmoothed sections in Figures G.2a and b. (b) $M_{\mathrm{NIP}}^{(\xi)}$ values extracted from the smoothed sections in Figures G.2e and f.

- Sort the remaining attribute values within the window by magnitude (separately for each attribute type). Take the mean of a specified fraction of attribute values around the median value and assign it to the considered central zero-offset sample location. If there are no remaining attributes, use the original attribute values of the central zero-offset sample.

- Repeat the procedure for each sample in the zero-offset section.

Figure G.1 illustrates the effect of the smoothing of the $\alpha$ and $R_{\mathrm{NIP}}$ sections on the stack results. It shows a detail from the CRS stack section of the real land dataset of Section 6.3. In Figure G.1a no attribute smoothing has been applied, while Figure G.1b shows the stack result with smoothed attributes. In this example, a smoothing window with dimensions $n_t = 7$ and $n_\xi = 11$ was used. The maximum allowed angle deviation was $\Delta\alpha = 1°$ and the coherence threshold was 0.02 (compare Figure 6.12b). The significant improvement in the stack result due to the attribute smoothing is obvious.

Figure G.2 shows details of the $\alpha$ and $R_{\mathrm{NIP}}$ sections from a different dataset with a low overall S/N ratio. The original attribute sections are displayed in Figures G.2a and b. Fluctuations and outliers in the attribute sections are clearly visible. Figures G.2c and d show the same sections, but with attributes associated with coherence values below a specified threshold masked. Figures G.2e and f show the attribute sections after smoothing (again, a mask has been applied). The fluctuations and outliers have clearly been removed. To illustrate the effect of the attribute smoothing on input data for the tomographic inversion, a number of $M_{\mathrm{NIP}}^{(\xi)}$ values extracted from the original, unsmoothed attribute sections (Figures G.2a and b) along two selected events are displayed in Figure G.3a. The corresponding $M_{\mathrm{NIP}}^{(\xi)}$ values extracted from the smoothed attribute sections (Figures G.2e and f) are shown in Figure G.3b. The data components exhibit a considerably improved continuity, which is physically more reasonable. They should, therefore, be much better suited to be used as input to the tomographic inversion than the values in Figure G.3a.

# List of Figures

# References

Aki, K. and Richards, P. G. (1980). *Quantitative Seismology – Theory and methods, Vol. I and II.* W. H. Freeman.

Al-Chalabi, M. (1973). Series approximation in velocity and traveltime computations. *Geophys. Prosp.*, 21:783–795.

Al-Yahya, K. (1989). Velocity analysis by iterative profile migration. *Geophysics*, 54(6):718–729.

Audebert, F. and Diet, J. P. (1993). Migrated focus panels: Focusing analysis reconciled with prestack depth migration. In *63rd Ann. Internat. Mtg., Expanded Abstracts*, pages 961–964. Soc. Expl. Geophys.

Audebert, F., Diet, J. P., Guillaume, P., Jones, I. F., and Zhang, X. (1997). CRP-scans: 3D PreSDM velocity analysis via zero-offset tomographic inversion. In *67th Ann. Internat. Mtg., Expanded Abstracts*, pages 1805–1808. Soc. Expl. Geophys.

Ben-Menahem, A. and Beydoun, W. B. (1985a). Range of validity of seismic ray and beam methods in general inhomogeneous media – I. general theory. *Geophys. J. R. astr. Soc.*, 82:207–234.

Ben-Menahem, A. and Beydoun, W. B. (1985b). Range of validity of seismic ray and beam methods in general inhomogeneous media – II. a canonical problem. *Geophys. J. R. astr. Soc.*, 82:235–262.

Berkovitch, A. and Gelchinsky, B. (1989). Inversion of common reflecting element (CRE) data (migration combined with interval velocity determination). In *59th Ann. Internat. Mtg., Expanded Abstracts*, pages 1250–1252. Soc. Expl. Geophys.

Berkovitch, A., Gelchinsky, B., and Keydar, S. (1994). Basic formulae for multifocusing stack. In *56th Mtg., Extended Abstracts*. Eur. Assn. Expl. Geophys.

Billette, F. and Lambaré, G. (1998). Velocity macro-model estimation from seismic reflection data by stereotomography. *Geophys. J. Int.*, 135:671–690.

Billette, F., Le Bégat, S., Podvin, P., and Lambaré, G. (2003). Practical aspects and applications of 2D stereotomography. *Geophysics*, 68(3):1008–1021.

Biloti, R., Santos, L. T., and Tygel, M. (2002). Multiparametric traveltime inversion. *Stud. geophys. geod.*, 46:177–192.

Bishop, T., Bube, K., Cutler, R., Langan, R., Love, P., Resnick, J., Shuey, R., Spindler, D., and Wyld, H. (1985). Tomographic determination of velocity and depth in laterally varying media. *Geophysics*, 50(1):903–923.

Bleistein, N. (1984). *Mathematical methods for wave phenomena*. Academic Press, Inc.

Bleistein, N. (1986). Two-and-one-half dimensional in-plane wave propagation. *Geophys. Prosp.*, 34:686–703.

Bleistein, N., Cohen, J. K., and Stockwell, Jr., J. W. (2001). *Mathematics of Multidimensional Seismic Imaging, Migration, and Inversion*. Interdisciplinary Applied Mathematics. Springer-Verlag.

Červený, V. (2001). *Seismic Ray Theory*. Cambridge University Press.

Červený, V. and Hron, F. (1980). The ray series method and dynamic ray tracing system for three-dimensional inhomogeneous media. *Bull. Seismol. Soc. Am.*, 70(1):47–77.

Chernyak, V. S. and Gritsenko, S. A. (1979). Interpretation of the effective common-depth-point parameters for a three-dimensional system of homogeneous layers with curvilinear boundaries. *Geologiya i Geofizika*, 20(12):112–120.

Claerbout, J. F. (1985). *Imaging the earth's interior*. Blackwell Scientific Publications.

Courant, R. and Hilbert, D. (1968). *Methoden der Mathematischen Physik I und II*. Springer-Verlag.

de Boor, C. (1978). *A practical guide to splines*. Springer-Verlag.

Delprat-Jannaud, F. and Lailly, P. (1993). Ill-posed and well-posed formulations of the reflection travel time tomography problem. *J. Geophys. Res.*, 98(B4):6589–6605.

Deregowski, S. M. (1986). What is DMO? *First Break*, 4(7):7–24.

Deregowski, S. M. (1990). Common-offset migrations and velocity analysis. *First Break*, 8(6):224–234.

Dix, C. H. (1955). Seismic velocities from surface measurements. *Geophysics*, 20(1):68–86.

Fagin, S. (1999). *Model-based depth imaging*. Soc. Expl. Geophys.

Farra, V. and Madariaga, R. (1987). Seismic waveform modeling in heterogeneous media by ray perturbation theory. *J. Geophys. Res.*, 92(B3):2697–2712.

Farra, V. and Madariaga, R. (1988). Non-linear reflection tomography. *Geophys. J.*, 95:135–147.

Faye, J. P. and Jeannot, J. P. (1986). Prestack migration velocities from focusing depth analysis. In *56th Ann. Internat. Mtg., Expanded Abstracts*, pages 438–440. Soc. Expl. Geophys.

Gardner, G. H. F., French, W. S., and Matzuk, T. (1974). Elements of migration and velocity analysis. *Geophysics*, 39(6):811–825.

Gazdag, J. (1978). Wave-equation migration by phase shift. *Geophysics*, 43:49–76.

Gilbert, F. and Backus, G. (1966). Propagator matrices in elastic wave and vibration problems. *Geophysics*, 31(2):326–332.

Gill, P. E., Murray, W., and Wright, M. H. (1981). *Practical Optimization*. Academic Press.

Gjøystdal, H., Reinhardsen, J. E., and Ursin, B. (1984). Traveltime and wavefront curvature calculations in three-dimensional inhomogeneous layered media with curved interfaces. *Geophysics*, 49(9):1466–1494.

Gjøystdal, H. and Ursin, B. (1981). Inversion of reflection times in three dimensions. *Geophysics*, 46(7):972–983.

Hertweck, T. (2004). *True-amplitude Kirchhoff migration: analytical and theoretical considerations*. Logos Verlag Berlin.

Höcht, G. (2002). *Traveltime approximations for 2D and 3D media and kinematic wavefield attributes*. PhD thesis, University of Karlsruhe.

Höcht, G., de Bazelaire, E., Majer, P., and Hubral, P. (1999). Seismics and optics: hyperbolae and curvatures. *J. Appl. Geoph.*, 42(3,4):261–281.

Hubral, P. (1983). Computing true amplitude reflections in a laterally inhomogeneous earth. *Geophysics*, 48(8):1051–1062.

Hubral, P. and Krey, T. (1980). *Interval velocities from seismic reflection traveltime measurements*. Soc. Expl. Geophys.

Jäger, R., Mann, J., Höcht, G., and Hubral, P. (2001). Common-reflection-surface stack: Image and attributes. *Geophysics*, 66(1):97–109.

Jannane, M., Beydoun, W., Crase, E., Cao, D., Koren, Z., Landa, E., Mendes, M., Pica, A., Noble, M., Roeth, G., Singh, S., Snieder, R., Tarantola, A., Trezeguet, D., and Xie, M. (1989). Wavelengths of earth structures that can be resolved from seismic reflection data. *Geophysics*, 54(7):906–910.

Jin, S. and Madariaga, R. (1994). Nonlinear velocity inversion by a two-step Monte Carlo method. *Geophysics*, 59(4):577–590.

Keydar, S., Edry, D., Berkovitch, A., and Gelchinsky, B. (1995). Construction of kinematic seismic model by homeomorphic imaging method. In *57th Mtg., Extended Abstracts*. Eur. Assn. Expl. Geophys.

Kosloff, D., Sherwood, J., Koren, Z., Machet, E., and Falkovitz, Y. (1996). Velocity and interface depth determination by tomography of depth migrated gathers. *Geophysics*, 61(5):1511–1523.

Kravtsov, Y. A. and Orlov, Y. I. (1990). *Geometrical optics of inhomogeneous media*. Wave Phenomena. Springer-Verlag.

Lafond, C. F. and Levander, A. R. (1993). Migration moveout analysis and depth focusing. *Geophysics*, 58(1):91–100.

Lailly, P. and Sinoquet, D. (1996). Smooth velocity models in reflection tomography for imaging complex geological structures. *Geophys. J. Int.*, 124:349–362.

Lanczos, C. (1961). *Linear differential operators*. Van Nostrand-Reinhold, Princeton.

Landa, E., Gurevich, B., Keydar, S., and Trachtman, P. (1999). Application of multifocusing method for subsurface imaging. *J. Appl. Geoph.*, 42(3,4):301–318.

Landa, E., Kosloff, D., Keydar, S., Koren, Z., and Reshef, M. (1988). A method for determination of velocity and depth from seismic reflection data. *Geophys. Prosp.*, 36:223–243.

Levenberg, K. (1944). A method for the solution of certain nonlinear problems in least squares. *Quaterly of Applied Mathematics*, 2:164–186.

Lines, L. R. and Treitel, S. (1984). Tutorial: A review of least-squares inversion and its application to geophysical problems. *Geophys. Prosp.*, 32:159–186.

Liu, Z. (1997). An analytical approach to migration velocity analysis. *Geophysics*, 62(4):1238–1249.

MacKay, S. and Abma, R. (1992). Imaging and velocity estimation with depth-focusing analysis. *Geophysics*, 57(12):1608–1622.

MacKay, S. and Abma, R. (1993). Depth-focusing analysis using a wavefront-curvature criterion. *Geophysics*, 58(8):1148–1156.

Mann, J. (2002). *Extensions and Applications of the Common-Reflection-Surface Stack Method*. Logos Verlag Berlin.

Mann, J. and Höcht, G. (2003). Pulse stretch effects in the context of data-driven imaging methods. In *65th Mtg., Expanded Abstracts*. Eur. Assn. Geosci. Eng.

Mann, J., Jäger, R., Müller, T., Höcht, G., and Hubral, P. (1999). Common-reflection-surface stack - a real data example. *J. Appl. Geoph.*, 42(3,4):283–300.

Marquardt, D. W. (1963). An algorithm for least squares estimation of non-linear parameters. *Journal of the Society of Industrial and Applied Mathematics*, 11:431–441.

Mayne, W. H. (1962). Common-reflection-point horizontal data stacking techniques. *Geophysics*, 27:927–938.

Menke, W. (1984). *Geophysical Data Analysis: Discrete Inverse Theory*. Academic Press.

Nemeth, T., Normark, E., and Qin, F. (1997). Dynamic smoothing in crosswell traveltime tomography. *Geophysics*, 62:168–176.

Nolet, G. (1987). Seismic wave propagation and seismic tomography. In Nolet, G., editor, *Seismic Tomography with Applications in Global Seismology and Exploration Geophysics*, pages 1–23. Reidel Publishing.

Oevel, W. (1996). *Einführung in die numerische Mathematik*. Spektrum Akademischer Verlag.

Ory, J. and Pratt, R. G. (1995). Are our parameter estimators biased? The significance of finite-difference regularization operators. *Inverse Problems*, 11:397–424.

Paige, C. C. and Saunders, M. A. (1982a). Algorithm 583 – LSQR: Sparse linear equations and least squares problems. *ACM Trans. Math. Softw.*, 8(2):195–209.

Paige, C. C. and Saunders, M. A. (1982b). LSQR: An algorithm for sparse linear equations and sparse least squares. *ACM Trans. Math. Softw.*, 8(1):43–71.

Popov, M. M. (2002). *Ray theory and Gaussian Beam method for Geophysicists*. Editora da Universidade Federal da Bahia.

Popov, M. M. and Pšenčík, I. (1978). Computation of ray amplitudes in inhomogeneous media with curved interfaces. *Studia Geoph. et Geod.*, 22:248–258.

Press, W. H., Teukolsky, S. A., Vetterling, W. T., and Flannery, B. P. (1992). *Numerical Recipes in C; The Art of Scientific Computing*. Cambridge University Press, 2nd edition.

Sattlegger, J. W., Rohde, J., Egbers, H., Dohr, G. P., Stiller, P. K., and Echterhoff, J. A. (1981). INMOD— two dimensional inverse modeling algorithm based on ray theory. *Geophys. Prosp.*, 29:229–240.

Schleicher, J., Tygel, M., and Hubral, P. (1993). Parabolic and hyperbolic paraxial two-point traveltimes in 3D media. *Geophys. Prosp.*, 41:495–513.

Schneider, W. (1978). Integral formulation for migration in two and three dimensions. *Geophysics*, 43:49–76.

Sen, M. and Stoffa, P. L. (1995). *Global Optimization Methods in Geophysical Inversion*. Elsevier, Amsterdam.

Stoffa, P. L., Fokkema, J. T., de Luna Freire, R. M., and Kessinger, W. P. (1990). Split-step Fourier migration. *Geophysics*, 55(04):410–421.

Stork, C. (1992). Reflection tomography in the postmigrated domain. *Geophysics*, 57(5):680–692.

Stork, C. and Clayton, R. W. (1991). Linear aspects of tomographic velocity analysis. *Geophysics*, 56(4):483–495.

Taner, M. T. and Koehler, F. (1969). Velocity spectra - digital computer derivation and applications of velocity functions. *Geophysics*, 34(06):859–881.

Tarantola, A. (1987). *Inverse Problem Theory: Methods for Data Fitting and Model Parameter Estimation*. Elsevier, Amsterdam.

Trappe, H., Gierse, G., and Pruessmann, J. (2001). Case studies show potential of Common Reflection Surface stack – structural resolution in the time domain beyond the conventional NMO/DMO stack. *First Break*, 19:625–633.

Ursin, B. (1982). Quadratic wavefront and traveltime approximations in inhomogeneous layered media with curved interfaces. *Geophysics*, 47(07):1012–1021.

van der Sluis, A. and van der Horst, H. A. (1987). Numerical solution of large, sparse linear algebraic systems arising from tomographic problems. In Nolet, G., editor, *Seismic Tomography with Applications in Global Seismology and Exploration Geophysics*, pages 49–83. Reidel Publishing.

Versteeg, R. J. (1993). Sensitivity of prestack depth migration to the velocity model. *Geophysics*, 58(6):873–882.

Williamson, P. R. (1990). Tomographic inversion in reflection seismology. *Geophys. J. Int.*, 100:255–274.

Woodward, M., Farmer, P., Nichols, D., and Charles, S. (1998). Automated 3D tomographic velocity analysis of residual moveout in prestack migrated common image point gathers. In *68th Ann. Internat. Mtg., Expanded Abstracts*, pages 1218–1221. Soc. Expl. Geophys.

Yilmaz, Ö. (1987). *Seismic Data Processing*. Soc. Expl. Geophys.

Yilmaz, Ö. (2001). *Seismic Data Analysis, Vols. 1 and 2*. Soc. Expl. Geophys.

Yilmaz, Ö. and Chambers, R. (1984). Migration velocity analysis by wavefield extrapolation. *Geophysics*, 49(10):1664–1674.

# Used hard- and software

The tomographic inversion program developed in the course of this thesis has been implemented in C++. It has been installed and applied on PCs running the operating system Linux and on an SGI Origin 3200 running IRIX. All 1D, 2D, and 3D inversion examples presented in the thesis were obtained using this program. It was also used for creating the synthetic test data in 2D and 3D smooth models.

The CRS stack results shown in this thesis were obtained using the implementation by Mann (2002), while for Kirchhoff depth migration, the program Uni3D (Hertweck, 2004) was used. Additional seismic processing was performed with the freely available seismic processing package Seismic Un*x (Center for Wave Phenomena, Colorado School of Mines) which was also used for the visualization of results (seismic data and velocity models). The synthetic seismic data used in this thesis were created with the Norsar 3D ray modeling tool.

For data visualization, the mathematical program package Matlab (The MathWorks) was used. Additional figures in the thesis were created using Maple (Maplesoft) and Xfig.

The thesis itself was written on a PC with the free operating system Linux (SuSE Linux 8.0), using the document preparation system LATEX, based on the typesetting system TEX.

# Danksagung

Zum Entstehen der vorliegenden Arbeit haben eine Reihe von Personen in verschiedener Weise beigetragen. Ihnen möchte ich an dieser Stelle herzlich danken.

Mein besonderer Dank gilt **Prof. Dr. Peter Hubral**, für die Betreuung der Arbeit. Er hat mich während meiner gesamten Zeit in Karlsruhe in jeder Hinsicht sehr unterstützt und mir die Möglichkeit und den Freiraum gegeben, in meiner Arbeit ohne äußere Zwänge meinen Ideen und Interessen nachzugehen. Zudem hat er mir die Teilnahme an zahlreichen internationalen Tagungen ermöglicht, wovon ich sehr profitiert habe.

**Prof. Dr. Friedemann Wenzel** möchte ich für die Übernahme des Korreferats danken.

Ich möchte mich außerdem bei allen jetzigen und ehemaligen Kollegen und Studenten aus der Arbeitsgruppe Reflexionsseismik ganz herzlich für das gute Arbeitsklima bedanken.

Ganz besonders danke ich **Dr. Jürgen Mann** und **Steffen Bergler**, dass ich mit ihnen ein Büro teilen durfte. Mit beiden habe ich gerne zusammengearbeitet und sehr von ihrem Wissen profitiert. Gemeinsam haben wir ausserdem auf diversen Reisen viel Spaß gehabt. Gleiches gilt für **Dr. Thomas Hertweck**. Auch von ihm habe ich sehr viel gelernt und bei verschiedensten Problemen (nicht nur in Computerfragen) von seiner Hilfsbereitschaft profitiert. Seine gründlichen Korrekturen haben außerdem sehr zur Verbesserung der vorliegenden Arbeit beigetragen.

Für die gute Zusammenarbeit möchte ich auch **Christoph Jäger** danken, sowie **Tilman Klüver**, der die Arbeit an der tomographischen Inversion mit CRS Attributen weiterführen wird, **Paola Chávez Zander**, deren Erfahrungen bei der Anwendung der Tomographie auf Realdaten zu Verbesserungen des Programms geführt haben und **Zeno Heilmann**, mit dem ich bei der Bearbeitung der Realdaten in Kapitel 6 zusammengearbeitet habe.

Auch allen anderen Mitarbeitern des Geophysikalischen Instituts, besonders **Claudia Payne** und **Petra Knopf**, sei an dieser Stelle für ihre Hilfe bei verschiedenen Gelegenheiten gedankt.

Für die unterhaltsamen gemeinsamen Mittagspausen möchte ich mich bei **Wolfgang Wirth**, **Michael Martin** und **Maren Böse** bedanken.

**Thomas**, **Jürgen**, **Christoph**, **Maren** und **Steffen** haben diese Arbeit korrekturgelesen, wofür ich mich herzlich bei ihnen bedanke.

Den Sponsoren des Wave Inversion Technology (WIT) Konsortiums möchte ich für ihre Unterstützung danken. Besonders **Dr Jürgen Pruessmann** und seinen Kollegen bei TEEC, Isernhagen, sowie **Paolo Marchetti** (ENI E&P, Milano) danke ich für ihr Interesse an meiner Arbeit und für ihre Hilfe bei der Einführung des im Rahmen dieser Arbeit entwickelten Programms in die professionelle Anwendung. Der Firma Hotrock EWK Offenbach/Pfalz GmbH danke ich für die Erlaubnis, die seismischen Daten in Kapitel 6 zu zeigen.

Mein ganz besonderer Dank gilt **Heike Becherer**, bei der ich auch in stressigen Phasen immer wieder abschalten und neue Energie tanken konnte.

Schließlich möchte ich mich herzlich bei meinen Eltern **Maja** und **Uwe Duveneck** bedanken, die mich immer in jeder Hinsicht unterstützt haben und mir das Studium der Geophysik ermöglicht haben.

# Lebenslauf

## Persönliche Daten

| | |
|---|---|
| Name: | Eric Duveneck |
| Geburtsdatum: | 8. August 1972 |
| Geburtsort: | Bremen |
| Nationalität: | deutsch |

## Schulausbildung

| | |
|---|---|
| 1979 - 1983 | Grundschule Annenheide, Delmenhorst |
| 1983 - 1985 | Orientierungsstufe, Schulzentrum Süd, Delmenhorst |
| 1989 - 1990 | Niceville High School, Florida, USA |
| 1985 - 1989 & 1990 - 1993 | Gymnasium an der Max-Planck-Straße, Delmenhorst |
| 13.05.1993 | Abitur |

## Zivildienst

| | |
|---|---|
| 08/1993 - 10/1994 | Gemeinde Unser Lieben Frauen, Bremen |

## Hochschulausbildung

| | |
|---|---|
| 1994 - 2000 | Studium der Geophysik an der Christian-Albrechts-Universität, Kiel |
| 10/1997 - 06/1998 | University of Southampton, England |
| 20.09.2000 | Diplom |
| seit 2001 | Doktorand an der Fakultät für Physik der Universität Karlsruhe (TH) |